Palgrave Studies in Digital Business & Enabling Technologies

Series Editors
Theo Lynn
Irish Institute of Digital Business
DCU Business School
Dublin, Ireland

John G. Mooney
Graziadio Business School
Pepperdine University
Malibu, CA, USA

This multi-disciplinary series will provide a comprehensive and coherent account of cloud computing, social media, mobile, big data, and other enabling technologies that are transforming how society operates and how people interact with each other. Each publication in the series will focus on a discrete but critical topic within business and computer science, covering existing research alongside cutting edge ideas. Volumes will be written by field experts on topics such as cloud migration, measuring the business value of the cloud, trust and data protection, fintech, and the Internet of Things. Each book has global reach and is relevant to faculty, researchers and students in digital business and computer science with an interest in the decisions and enabling technologies shaping society.

More information about this series at
http://www.palgrave.com/gp/series/16004

Theo Lynn • John G. Mooney
Brian Lee • Patricia Takako Endo
Editors

The Cloud-to-Thing Continuum

Opportunities and Challenges in Cloud,
Fog and Edge Computing

Editors
Theo Lynn
Irish Institute of Digital Business
DCU Business School
Dublin, Ireland

John G. Mooney
Graziadio Business School
Pepperdine University
Malibu, CA, USA

Brian Lee
Software Research Institute
Athlone Institute of Technology
Dublin, Ireland

Patricia Takako Endo
Irish Centre for Cloud Computing and
Commerce
DCU Business School
Recife, Brazil

ISSN 2662-1282   ISSN 2662-1290   (electronic)
Palgrave Studies in Digital Business & Enabling Technologies
ISBN 978-3-030-41109-1   ISBN 978-3-030-41110-7   (eBook)
https://doi.org/10.1007/978-3-030-41110-7

PREFACE

This is the fifth book in the series, *Advances in Digital Business and Enabling Technologies*, which aims to contribute to multi-disciplinary research on digital business and enabling technologies, such as cloud computing, social media, big data analytics, mobile technologies and the Internet of Things, in Europe.

Previous volumes sought to consider and extend conventional thinking on disrupting finance, the business value of the cloud, and more specifically on cloud computing architectures to greater support heterogeneity and to reliably provision applications and infrastructure on distributed clouds. In many ways, this fifth volume is a companion to the last volume, *Managing Distributed Cloud Applications and Infrastructure*. Whereas that volume focussed on optimisation from the cloud to the edge, this fifth volume explores the cloud-to-thing continuum.

The 'Network Nation', the 'Wired Society', and 'the Networked Society' are just some of the terms used over the last forty years to describe communities organised around systems of interaction, systems of resource allocation, and systems of integration and co-ordination driven by advances in information and communications technologies (Martin 1978; Hiltz and Turoff 1978). In the last decade, we have seen the emergence of new terms, the 'third ICT platform' and the 'fourth industrial revolution' (4IR) characterized by the ubiquity, convergence and interdependence of next generation technologies—social media, mobile, cloud, big data, and sensor technologies—with the promised of transforming how society operates and interacts (IDC 2013). However, we are not there yet. The

Internet of Things represents a significant first step towards the networked society. It offers massive societal and economic opportunities while at the same time significant challenges not least the delivery and management of the technical infrastructure underpinning it and the deluge of data generated from it, ensuring privacy and security, and capturing value from it. This book explores these challenge, presenting the state of the art and future directions for research but also frameworks for making sense of this complex area.

The content of the book is based on contributions from researchers on the RECAP project, a European Union project funded under Horizon 2020 [recap-project.eu] but also collaborators from SFI CONFIRM Centre in Ireland and UFPE, Brazil.

Chapter 1 defines the Internet of Things and introduces key concepts and enabling technologies. It provides a sense-making framework that marries technical and socio-technical perspectives and summarises some of the main Cloud-IoT reference architectures.

Chapter 2 revisits conventional cloud computing and discusses how cloud computing is evolving as a result of the Internet of Things. It discusses how new processor architectures and service models are changing the essence of what we think of as conventional cloud computing. But more than that, Chap. 2 explores how cloud computing is moving from being a centralised cloud to a distributed one and from being a homogenous cloud to a heterogeneous one. This presents the opportunity for new approaches for resource provisioning, self-organisation and self-management, and delivering a separation of concerns, all critical for the future of a cloud capable of supporting the Internet of Things.

Chapter 3 tracks the evolution of 5G network technologies. While innovations such as Ipv6 and new paradigms in computing such as fog, edge and dew computing are enabling the IoT, LTE and 5G play a critical role in network connectivity. This chapter explains why RAN designs are critical to 5G success and consequently, the success of the Internet of Things.

Chapter 4 review the state of the art with regards to orchestration along the cloud-to-thing continuum with a specific emphasis on container-based orchestration and fog-specific orchestration architectures. The effective management of complex and heterogeneous computing environments is one of the biggest challenges that service and infrastructure providers are facing in the cloud-to-thing continuum era. This chapter highlights the need for fog-specific standards and orchestrators to accelerate momentum in the same way that cloud-native applications gave momentum to the development of cloud orchestrators.

Chapter 5 discusses some of the challenges in high-criticality Internet of Things use cases. Storing and processing at the end device (the edge), at the intermediary layer (the fog), or centrally (the cloud) introduces new points of potential failure. For high-criticality use cases, any downtime impacting one or more components in the architecture can result in adverse effects and/or additional logistical effort and cost. This chapter discusses extant research on how cloud, fog and edge computing is being used in smart city, smart agriculture and e-health systems.

Chapter 6 explores security issues in edge computing with a particular emphasis on distributed intelligence. Due to resource-constrained hardware and software heterogeneities, most edge computing systems are prone to a large variety of attacks. Incorporating intelligence in edge computing systems leads to new security vulnerabilities including data and model poisoning, and evasion attacks. This chapter presents a discussion on the most pertinent threats to edge intelligence and countermeasures to deal with the threats.

Chapter 7 explores privacy in the Internet of Things from the perspective of consumers. This chapter discuses data privacy and trust research on the Internet of Things and posits that to foster a sense of privacy and trust among consumers, IoT service providers must communicate with consumers regarding their data practices in a transparent manner. This chapter proposes an Internet of Things Trust Label and present a framework for testing the effectiveness of privacy disclosures in building consumers' perceptions of privacy and trust and empowering consumers to adopt IoT devices whilst retaining some level of privacy.

Chapter 8 presents a general framework for mapping the business value of investments in the Internet of Things which aims to support managers in their decision-making process by providing an overview of how specific resources needs to be linked together in order to generate business value. The presented framework is also used as a point of reference for identifying current research gaps which may represent avenues for future research.

Dublin, Ireland   Theo Lynn
Malibu, CA   John G. Mooney
Dublin, Ireland   Brian Lee
Recife, Brazil   Patricia Takako Endo

REFERENCES

Hiltz, S. R., and Turoff, M. 1993. *The Network Nation: Human Communication via Computer.* MIT Press.

IDC. 2013. *The 3rd Platform: Enabling Digital Transformation.* USA: IDC.

Martin, J. 1978. *The Wired Society.* NJ: Prentice-Hall.

ACKNOWLEDGEMENT

This book funded by the European Union's Horizon 2020 Research and Innovation Programme through the RECAP project (https://recap-project.eu) under Grant Agreement Number 732667. Research in this book was partially funded with the financial support of Science Foundation Ireland (SFI) under Grant Number SFI 16/RC/3918, co-funded by the European Regional Development Fund, and the World Technology Universities Network.

CONTENTS

NOTES ON CONTRIBUTORS

Saeed H. Alsamhi is an Assistant Professor in IBB University, Ibb, Yemen and Assistant Professor in Shenzhen Institute of Advanced Technology, Shenzhen, China. His research interests include distributed intelligence, edge computing, AI, the Internet of Things, smart healthcare, robot collaboration, drone technology, and smart cities.

Mohammad S. Ansari is a postdoctoral researcher at Software Research Institute, AIT, Ireland, and an Assistant Professor at Aligarh Muslim University, India. His research interests include neural networks, machine learning, edge computing and smart cities. His articles have appeared in over 100 publications.

Gibson B. N. Barbosa is a doctoral student at the Universidade Federal de Pernambuco. He has experience in computational intelligence, clustering analysis, integrated circuit design through prototyping, FPGA, and testing and verification of RTL systems.

Malika Bendechache is a postdoctoral researcher at the Irish Institute of Digital Business at Dublin City University. She received her PhD in Computer Science at University College Dublin (UCD) in the area of parallel and distributed data mining. Dr Bendechache was previously a researcher at the Insight Centre for Data Analytics at UCD. Her research interests span across distributed systems, big data analytics, and simulation of large-scale distributed cloud, fog, and edge computing environments and associated applications.

Lubnnia M. F. de Souza has a Masters in Computer Science from the Universidade Federal de Pernambuco. Her research interests include performance and dependability evaluation, Petri Nets, formal models, business process modelling and evaluation, the Internet of Things, and smart cities.

Jörg Domaschka is a Senior Researcher and Group Manager at the Institute of Information Resource Management at the University of Ulm. He holds a Diploma in Computer Science from FAU, Erlangen Nuremberg, and a PhD in Computer Science from the University of Ulm. His research interests include distributed systems, fault-tolerance, middleware platforms, and NoSQL databases. Current focus of his work lies on middleware and run-time systems for geo-distributed infrastructure and applications.

Patricia Takako Endo is a postdoctoral research fellow at Irish Institute of Digital Business, Dublin City University, Ireland, a Professor at Universidade de Pernambuco, Brazil, and a researcher at Grupo de Pesquisa em Redes e Telecomunicações (GPRT). Her research interests include cloud computing, fog computing, Internet of Things, system availability, and data analytics. Her articles have appeared in over 110 publications in the above research areas.

Leylane Ferreira is a postgraduate student in Computer Science at Universidade Federal de Pernambuco and a researcher at Grupo de Pesquisa em Redes e Telecomunicações (GPRT). Her current research interests include cloud and fog computing, and optimization algorithms.

Christos K. Filelis Papadopoulos received his Diploma in Engineering degree from the Electrical and Computer Engineering Department of the Democritus University of Thrace, Greece, in 2010 and his Ph.D. in Numerical Analysis and High Performance Scientific Computations from the same university in 2014. His research interests include preconditioned iterative methods, multigrid and multilevel methods as well as parallel computing.

Grace Fox is an Assistant Professor of Digital Business at Dublin City University Business School. Her research interests intersect the broad interdisciplinary areas of information privacy and digital technology adoption and assimilation. Her research has been published in premier academic journals such as Information Systems Journal and Communications of the Association of Information Systems along with numerous peer-ranked chapters and International conferences in the management, information systems and computer science domains.

Konstantinos M. Giannoutakis is a Postdoctoral Research Fellow at the Information Technologies Institute of Centre for Research and Technology Hellas. His research interests include high performance and scientific computing, parallel systems, grid/cloud computing, service-oriented architectures and software engineering techniques. His articles have appeared in over 80 publications in the above research areas.

Glauco E. Gonçalves is a Professor at the Rural Federal University of Pernambuco (UFRPE). He received his Ph.D. in Computer Science from UFPE in 2012. His research interests include cloud computing, the Internet of Things, system modeling and optimization.

Frank Griesinger is a researcher at the Institute for Organisation and Management of Information systems (OMI) at the University of Ulm, Germany. He has been involved in the EC funded research projects RECAP, PaaSage and CloudSocket. His research interests include cloud computing, IoT, description languages for cloud-native applications, and their execution in a highly distributed manner.

Judith Kelner received her PhD from the Computing Laboratory at the University of Kent in Canterbury, UK in 1993. She has been a Full Professor at Universidade Federal de Pernambuco since 1979. Currently, she leads the GRVM team as well as coordinates a number of research projects in the areas of multimedia systems, design of virtual and augmented reality applications, and smart communication devices.

Brian Lee received his Ph.D. from Trinity College Dublin. He worked in the Telecommunications industry for many years in network management research development. He is currently the Director of the Software Research Institute at Athlone IT. His research interests are centred on the broad theme of 'responsive infrastructures' across the areas of computer security and networking.

Theo Lynn is Full Professor of Digital Business at Dublin City University and is Director of the Irish Institute of Digital Business. He was formerly the Principal Investigator (PI) of the Irish Centre for Cloud Computing and Commerce, an Enterprise Ireland/IDA-funded Cloud Computing Technology Centre. Professor Lynn specialises in the role of digital technologies in transforming business processes with a specific focus on cloud computing, social media and data science.

Kayo H. de C. Monteiro is a postgraduate student at the Programa de Pós-Graduação em Engenharia da Computação (PPGEC) at Universidade de Pernambuco. He holds a postgraduate degree in Strategic Information Technology Management from Universidade Estácio de Sá at an undergraduate degree in Information Systems from Universidade de Pernambuco. His research interests include computational intelligence, cloud and fog computing, distributed systems.

John G. Mooney is Associate Professor of Information Systems and Technology Management and Academic Director of the Executive Doctorate in Business Administration at the Graziadio Business School. Dr. Mooney previously served as Executive Director of the Institute for Entertainment, Media and Culture from 2015 to 2018. He was named Fellow of the Association for Information Systems in December 2018. His current research interests include management of digital innovation (i.e. IT-enabled business innovation) and business executive responsibilities for managing digital platforms and information resources.

André L. C. Moreira received his Ph.D. in Computer Science from the Universidade Federal de Pernambuco. His research topic was in self organization of cloud networks and adaptation of CDN provisioning algorithms.

Yuansong Qiao is a Senior Research Fellow in the Software Research Institute at Athlone Institute of Technology, Ireland. He received his Ph.D. in Computer Applied Technology from the Institute of Software, Chinese Academy of Sciences. His current research interests include Future Internet Architecture, Blockchain Systems, IoT Systems, Smart Manufacturing and Edge Intelligence and Computing.

Andrea Maria N. C. Ribeiro is a doctoral student at Universidade Federal de Pernambuco (UFPE). She is currently a researcher at Grupo de Pesquisa em Redes e Telecomunicações (GPRT). Her fields of interest are computer networks, telecommunications, signal processing, energy efficiency, sensors and the Internet of Things.

Élisson da S. Rocha is a postgraduate student of the Programa de Pós-Graduação em Engenharia da Computação (PPGEC) at the Universidade de Pernambuco where he graduated with a Bachelor's degree in in Information Systems. He is a researcher at the Universidade Federal de

Pernambuco Grupo de Pesquisa em Redes e Telecomunicações (GPRT). His research interests include computational intelligence, neural networks, cloud computing, and distributed systems.

Pierangelo Rosati is Assistant Professor in Business Analytics at DCU Business School and Director of Industry Engagement at the Irish Institute of Digital Business. Dr. Rosati holds a PhD in Accounting and Finance from the University of Chieti-Pescara (Italy) and an MSc in Management and Business Administration from the University of Bologna. He was appointed Visiting Professor at the University of Edinburgh Business School, Universidad de las Américas Puebla and at Católica Porto Business School, and visiting Ph.D. Student at the Capital Markets Cooperative Research Center (CMCRC) in Sydney. Dr. Rosati has been working on research projects on FinTech, Blockchain, cloud computing, data analytics, business value of IT, and cyber security.

Djamel Sadok received his PhD in Computer Science at the University of Kent in Canterbury, UK in 1990. He has been a member of staff at Universidade Federal de Pernambuco since 1993. His research interests include communication systems, access networks, security, cloud computing and traffic classification. Currently, he leads the Grupo de Pesquisa em Redes e Telecomunicações (GPRT) team at UFPE.

Guto L. Santos is a doctoral student at Universidade Federal de Pernambuco and a researcher at the Grupo de Pesquisa em Redes e Telecomunicações (GPRT). His research interests include cloud computing, the Internet of Things, fog computing, 5G, and machine learning including deep learning.

Minas Spanopoulos-Karalexidis is a research assistant at the Information Technologies Institute of Centre for Research and Technology Hellas. His research interests include high performance scientific computing, simulation methods, sparse matrix technologies, iterative methods, parallel and distributed systems and static timing analysis.

Sergej Svorobej is a postdoctoral researcher in the Irish Institute of Digital Business at Dublin City University. Sergej's research focus is on complex systems, modelling and simulation with specific emphasis on cloud computing applications and infrastructure. Prior to working on the Horizon 2020 RECAP project, Sergej was a researcher at the Irish Centre for Cloud Computing and Commerce and on the FP7 CACTOS project. Previously, he held roles in SAP Ireland and SAP UK. He holds a Ph.D. from

Dublin City University and a B.Sc. in Information Systems and Information Technology from Dublin Institute of Technology.

Dimitrios Tzovaras is the Director (and Senior Researcher Grade A') of the Information Technologies Institute. He received a Diploma in Electrical Engineering and a Ph.D. in 2D and 3D Image Compression from the Aristotle University of Thessaloniki, Greece in 1992 and 1997, respectively. Prior to his current position, he was Senior Researcher on the Information Processing Laboratory at the Electrical and Computer Engineering Department of the Aristotle University of Thessaloniki. His main research interests include network and visual analytics for network security, computer security, data fusion, biometric security, virtual reality, machine learning and artificial intelligence.

Yuhang Ye received his Ph.D. from Athlone Institute of Technology, Ireland in the application of multipath congestion control for the future Internet. He received his Bachelor of Engineering degree in Electronic Engineering from National University of Ireland, Maynooth, Ireland and Bachelor of Science degree in Information Technology from Changzhou University, China in 2012. Dr. Ye is currently a post-doctoral researcher in Software Research Institute in Athlone Institute of Technology. His current research interests include network security, blockchain and price-based congestion control.

ABBREVIATIONS

5G	Fifth Generation
ADC	Analogue-to-Digital Conversion
AI	Artificial Intelligence
AIDC	Automatic Identification and Data Capture
API	Application Programming Interface
ARPU	Average Revenue per User
BBU	Baseband Unit
BS	Base Station
C2T	Cloud-to-Thing
CAM	Centralized Autonomic Managers
CapEx	Capital Expenditure
CBA	Cost-Benefit Analysis
CCTV	Closed Circuit Television
CNN	Convolutional Neural Network
CoMP	Coordinated Multipoint
CONCERT	Cloud-Based Architecture for Next-Generation Cellular Systems
COP	Control Orchestration Protocol
CPRI	Common Public Radio Interface
CPU	Central Processing Unity
C-RAN	Cloud Radio Access Network
CSRF	Cross-Site Request Forgery
D2D	Device-to-Device
DAS	Distributed Antenna Systems
DDoS	Distributed Denial-of-Service
DL	Deep Learning
DoS	Denial of Service (attack)
D-RAN	Distributed Radio Access Network

DRL	Deep Reinforcement Learning
DTN	Delay Tolerant Networking
DU	Digital Unit
EAT	Ensemble Adversarial Training
EB	ExaBytes
EPC	Electronic Product Code
ETSI	European Telecommunications Standards Institute
EVA	Economic Value Added
FA	Fog Agent
FaaS	Function-as-a-Service
F-AP	Fog Computing-Based Access Point
FL	Federated Learning
FO	Fog Orchestrator
FPGA	Field-Programmable Gate Arrays
F-RAN	Fog Radio Access Network
F-UE	Fog user equipment
Gbps	Gigabytes per Second
GDPR	General Data Protection Regulation
GE	General Electric
GHz	Gigahertz
GPU	Graphics Processing Unit
H-CRAN	Heterogeneous Cloud Radio Access Network
HetNet	Heterogeneous Networks
HPC	High Performance Computing
HPN	High Power Node
HTML	Hypertext Markup Language
HTTP	HyperText Transfer Protocol
IaaS	Infrastructure-as-a-Service
ICMP	Internet Control Message Protocol
IDE	Integrated Development Environment
IID	Independent and Identically Distributed
IIIRA	Industrial Internet Reference Architecture
IIoT	Industrial Internet of Things
ILP	Integer Linear Programming
IoC	Intelligent Operations Center
IoHT	Internet of Healthcare Things
IoMT	Internet of Medical Things
IoT	Internet of Things
IoT ARM	IoT Architectural Reference Model
IPv6	Internet Protocol version 6
IRA	IoT Reference Architecture
IRR	Internal Rate of Return

IS	Information System
KPI	Key Performance Indicator
LPN	Low Power Node
LTE	Long Term Evolution
M2M	Machine-to-Machine
MANO	Management and Orchestration
MBS	Macro cell base station
MIMO	Multiple-input, Multiple-output
ML	Machine Learning
NAS	Network-Aware Scheduling
NFV	Network Function Virtualisation
NFVI	Network Function Virtualisation Infrastructure
NPV	Net Present Value
OpEx	Operating Expenditure
OPNFV	Open Platform for NFV
OS	Operating System
OSM	Open Source Mano
P2PN	P2P Negotiation
PaaS	Platform-as-a-Service
PDIoT	Power Distribution IoT
PII	Personally Identifiable Information
PMT	Protection Motivation Theory
PoD	Ping of Death
QoE	Quality of Experience
QoS	Quality of Service
RA	Reference Architecture
RAN	Radio Access Network
RANaaS	RAN-as-a-Service
RBD	Reliability Block Diagram
RF	Radio Frequency
RFID	Radio Frequency Identification
ROI	Return on Investment
RONI	Reject on Negative Impact
RRH	Remote radio head
RU	Radio unit
SaaS	Software-as-a-Service
SAS	System Architecture Specifications
SBS	Small Cell Base Station
SCT	Social Contract Theory
SDK	Software Development Kit
SDN	Software Defined Networking
SGD	Stochastic Gradient Descent

SLA	Service-Level Agreements
SMC	Secure Multi-Party Computation
SOA	Service Oriented Architecture
SOAFI	Service Orchestration Architecture for Fog-enabled Infrastructures
SORTS	Supporting the Orchestration of Resilient and Trustworthy Fog Services
SOSM	Self-Organized Self-Managed
SPN	Stochastic Petri Net
SQL	Structured Query Language
SSRF	Server-Side Request Forgery
STRIP	STRong Intentional Perturbation
SVaaS	Smart-Vehicle-as-a-Service
SYN	Synchronization (packets)
TCO	Total Cost of Ownership
tRONI	Targeted Reject on Negative Impact
UDP	User Datagram Protocol
UI	User Interface
VIM	Virtualised Infrastructure Manager
VM	Virtual Machine
VNF	Virtual Network Function
VNFM	VNF Manager
VNFO	VNF Orchestrator
VTN	Virtual Tenant Networks
WNC	Wireless Network Cloud
WPA	Wi-Fi Protected Access
XML	Extensible Markup Language
XSS	Cross-site scripting

LIST OF FIGURES

LIST OF TABLES

CHAPTER 1

The Internet of Things: Definitions, Key Concepts, and Reference Architectures

Theo Lynn, Patricia Takako Endo,
Andrea Maria N. C. Ribeiro, Gibson B. N. Barbosa,
and Pierangelo Rosati

Abstract This chapter introduces the Internet of Things (IoT) and presents definitions and a general framework for conceptualising IoT. Key concepts and enabling technologies are summarised followed by a synthesis and discussion of the current state-of-the-art in IoT Reference Architectures.

Keywords Internet of things • IoT • IoT Reference Architecture

T. Lynn • P. Rosati
Irish Institute of Digital Business, DCU Business School, Dublin, Ireland
e-mail: theo.lynn@dcu.ie; pierangelo.rosati@dcu.ie

P. T. Endo
Irish Institute of Digital Business, Dublin City University, Dublin, Ireland

Universidade de Pernambuco, Recife, Brazil
e-mail: patricia.endo@upe.br

Andrea Maria N. C. Ribeiro • G. B. N. Barbosa (✉)
Universidade Federal de Pernambuco, Recife, Brazil
e-mail: andrea.maria@gprt.ufpe.br; gibson.nunes@gprt.ufpe.br

T. Lynn et al. (eds.), *The Cloud-to-Thing Continuum*, Palgrave
Studies in Digital Business & Enabling Technologies,
https://doi.org/10.1007/978-3-030-41110-7_1

1.1   Introduction

The Internet has evolved in a series of waves (Cisco 2012). The first three waves were device-centric. In the first wave, we went to a device, typically a desktop PC, to access the Internet. As mobile computing evolved, soon we brought our own devices with us and could access the Internet anywhere anytime. Today, we are in the midst of the so-called Internet of Things (IoT) where devices (things) are connected to the Internet and each other. These things comprise a multitude of heterogeneous devices ranging from consumer devices, such as mobile phones and wearables, to industrial sensors and actuators. Gartner (2017) estimated only 8.4 billion things were connected in 2017 representing just over 0.5% of the total estimated connectable physical objects worldwide.

This objective of this chapter is to introduce readers to the Internet of Things. The remainder of the chapter is organised as follows. First, we will explore perspectives on the definition of the Internet of Things (IoT) followed by key constructs and concepts underlying IoT including a general research framework for conceptualising IoT. Then, we will delve into a further level of granularity and present a selection of IoT Reference Architectures before concluding.

1.2   Defining the Internet of Things

The Internet of Things (IoT) has rapidly grown in prominence in the last ten years and, yet, it means different things to different people. Indeed Whitmore et al. (2015) note that there is no universal definition of IoT. Two main conceptualisations exist—the technical and socio-technical perspectives. The first, the pure technical perspective, views IoT as an assemblage and ecosystem of technical artefacts. It is defined by reference to these artefacts and their capabilities. These range in detail. For example, Weyrich and Ebert 2016, p. 1) define IoT as being *"[...] about innovative functionality and better productivity by seamlessly connecting devices."* In contrast, Tarkoma and Katasonov (2011, p. 2) is significantly more detailed defining IoT as a *"global network and service infrastructure of variable density and connectivity with self-configuring capabilities based on standard and interoperable protocols and formats [which] consists of heterogeneous things that have identities, physical and virtual attributes, and are seamlessly and securely integrated into the Internet."* Similarly, Whitmore et al. (2015, p. 1) define the IoT as *"a paradigm where everyday objects can*

*be equipped with identifying, sensing, networking and processing capabilities
that will allow them to communicate with one another and with other devices
and services over the Internet to achieve some objective.*" Unsurprisingly,
given the nature of these definitions, they dominate Computer Science
literature.

The socio-technical perspective of IoT recognises not only the technical artefacts but also the associate actors and processes with which the IoT
interacts. For example, Haller et al. (2009) recognises the role of the connected objects as active participants in business processes. They define the
IoT as "*a world where physical objects are seamlessly integrated into the
information network, and where the physical objects can become active participants in business processes. Services are available to interact with these
'smart objects' over the Internet, query their state and any information associated with them, taking into account security and privacy issues*" (Haller
et al. 2009, p. 15). Shin (2014, p. 25) argues that the IoT is part of
"*wider, socio-technical systems, comprising humans, human activity, spaces,
artefacts, tools and technologies.*" Indeed, Shin et al. note that in some
instances, a biological entity may, in fact, be considered the connected
thing, for example a human with a heart monitor implant or a farm animal
with a biochip transponder.

This perspective taken in this book is not particularly concerned with a
specific IoT-related definition or problem. Figure 1.1 below presents a

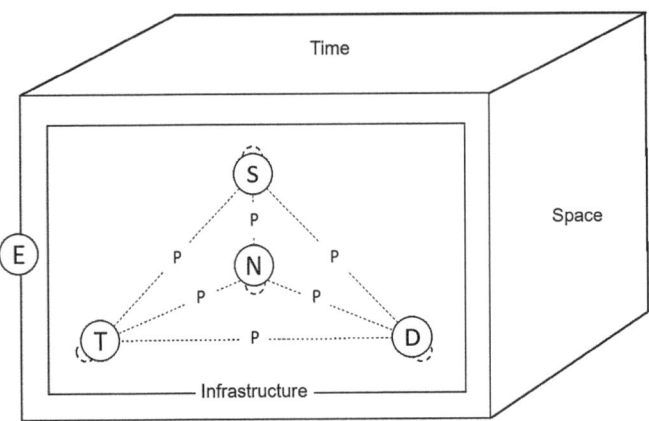

Fig. 1.1 A general framework for conceptualising big data research in the
Internet of Things

general research framework for conceptualising IoT research. It is general in that it is capable of being used to understand IoT related problems and research questions in conjunction with widely accepted levels of generalisation (abstraction) in both the social sciences (nano, micro, meso, macro) and computer sciences (computation, algorithmic/representational, physical/implementation). Furthermore, it provides a sufficiently general abstraction of the IoT in that it facilitates sense making without getting in to a non-generalisable level of granularity.

In this framework, five core entities are identified and defined—social actors, things, data, networks, and events. Each of these entities has a myriad of characteristics that may change and evolve over time and inflect our understanding of how value can be generated and captured at different units of analysis:

- *Social Actors (S)*, while typically human, need not be; the framework is flexible enough to accommodate the emerging concept of computers as social actors (Lynn et al. 2015; Zhao 2003).
- *Things (T)* are primarily physical however they may also be virtual and exist in augmented and/or virtual reality. Two key functional requirements of things in IoT and IoE are data sensing (collating data) and network connectivity.
- *Data (D)* here are discrete artefacts that can connect to other entities including other data and may be sourced from first party, second party, or third party sources. It recognises the existence of an IoT data chain. For example, Radio frequency identification (RFID) enables the tracking of objects through an electronic product code (EPC) serving serves as a link to data about the object that can queried over the Internet (Haller et al. 2009).
- *Networks (N)* are systems of interconnected entities and are both conduits and entities in themselves. Our framework accommodates networks between different types of IoT entities and those of the same type, for example machine-to-machine (M2M) networks.
- *Events (E)* are occurrences of interest at given time and/or physical or virtual space.
- *Processes (P)* are obviously critical to how entities interoperate in the IoT and comprise general (e.g. communication) and domain-specific processes. They are essential to how value is created, captured, and delivered in the IoT.

All entities and processes take place in an infrastructural setting and the framework recognises that in the IoT, additional data and metadata is created and collated at the infrastructural level. For example, depending on the networking, processing, and storage capabilities of a given device, these activities may be centralised (in the cloud), at the edge (at the device), or in an intermediary layer (the fog) and not only store or process this data but also may extract other hardware, software, functional use, or other ambient data that can provide different and/or new insights. Finally, each IoT use case is situated in space (physical or virtual) and time and it is against this context that different types of events occur and impact the IoT.

As the IoT can be explored from numerous perspectives, we argue that such a research framework can play an important role for researchers to make sense of a complex and dynamic environment and isolate the major constituents of the IoT experience. In addition, the proposed framework can be used as a general-purpose scaffold for crafting research agendas on the IoT and avoiding duplicated and unfocussed research endeavours.

1.3   Key Concepts and Constructs

IoT revolves around a number key concepts and enabling technologies including object (thing) identification (e.g. IPv6), information sensing (e.g. RFID, sensors, GPS, etc.), communications technologies for data exchange, and network integration technologies (Shin 2014).

It is important to note that legacy computing and telecommunications architectures were not designed with the IoT in mind. The scale of heterogeneous devices and an unprecedented volume, variety and velocity of data combined with an extreme variation in use context require new paradigms in computing. Depending on the use case and service level requirements, IoT devices may require processing and storage locally, in the cloud or somewhere in between. In addition cloud computing, edge, fog, and dew computing are three new computing paradigms designed to support IoT. While beyond the scope of this chapter, it is useful to be aware of these concepts and technologies when consider the architectures in Sect. 1.4. Table 1.1 provides a brief definition for technology.

Table 1.1 Definitions of key technologies in IoT

Construct	Definition
Cloud computing	A model for enabling ubiquitous, convenient, on-demand network access to a shared pool of configurable computing resources (e.g. networks, servers, storage, applications, and services) that can be rapidly provisioned and released with minimal management effort or service provider interaction (Mell and Grance 2011, p. 2).
Dew computing	Dew computing is an on-premises computer software-hardware organisation paradigm in the cloud computing environment where the on-premises computer provides functionality that is independent of cloud services and is also collaborative with cloud services (Wang 2016).
Edge computing	Edge computing is the network layer encompassing the end devices and their users, to provide, for example, local computing capability on a sensor, metering or some other devices that are network-accessible (adapted from Iorga et al. 2017).
Fog computing	Fog computing is a layered model for enabling ubiquitous access to a shared continuum of scalable computing resources. The model facilitates the deployment of distributed, latency-aware applications and services, and consists of fog nodes (physical or virtual), residing between smart end-devices and centralised (cloud) services (adapted from Iorga et al. 2017).
IPv6	Internet Protocol version 6 (IPv6) is the most recent version of the Internet Protocol (IP). It is an identification and location system for computers on networks and routes traffic across the Internet. It dramatically expands the addressing space (IPv6 2003) thus facilitating the identification of smart objects.
Machine-to-machine communication (M2M)	M2M communication technologies provide capabilities for devices to communicate with each other through wired and wireless systems (Tsai et al. 2012, p. 1).
Radio frequency identification (RFID)	RFID is a form of automatic identification and data capture (AIDC) technology that uses electric or magnetic fields at radio frequencies to transmit information. Each object that needs to be identified has a small object known as an RFID tag affixed to it or embedded within it. The tag has a unique identifier and may optionally hold additional information about the object. Devices known as RFID readers wirelessly communicate with the tags to identify the item connected to each tag and possibly read or update additional information stored on the tag. This communication can occur without optical line of sight and over greater distances than other AIDC technologies (Karygiannis et al. 2007, p. ES-1).
Wireless sensor and actuator networks (WSAN)	WSANs are networks of large numbers of minimal capacity sensing, computing, and communicating devices and various types of actuators (Stankovic 2008).

1.4   IoT Reference Architectures

IoT devices are being used in a wide range of domains such as health, agriculture, smart cities, and process automation. The 'things' used can be characterised by their heterogeneity in terms of computing resources (processing, memory, and storage), network connectivity (communication protocols and standards) and software development (high degree of distribution, parallelisation, dynamicity). While such heterogeneity enables the depth and breadth of applications and use cases, it also introduces complexity, particularly with respect to expected service level requirements, for example, user and device mobility, software dependability, high availability, scenario dynamicity, and scalability. As such, an abstraction layer to promote interoperability amongst IoT devices is needed. However, lack of standardisation means that such interoperability is lacking (Cavalcante et al. 2015). Reference Architectures can help IoT software developers to understand, compare, and evaluate different IoT solutions following a uniform practice.

Several Reference Architectures have been proposed in order to standardise concepts and implementation of IoT systems in different domains. Breivold (2017), for instance, conducted a comparative study with eleven different Reference Architectures. This chapter focuses on the those Reference Architectures that enable IoT integration with cloud computing and/or fog and edge computing i.e. across the cloud to thing (C2T) continuum. Figure 1.2 shows the timeline containing the main Reference Architectures that support IoT across the C2T continuum, namely IoT Architectural Reference Model (IoT ARM), IEEE P2413 (IEEE P2413 2014), Industrial Internet Reference Architecture (IIRA) (Lin et al. 2019), WSO2 IRA, Intel SAS, Azure IRA, and SAT-IoT.

Each of the architectures below can be explored through the lens of the framework presented in Sect. 1.2 and embodies the key concepts and constructs discussed in Sect. 1.3.

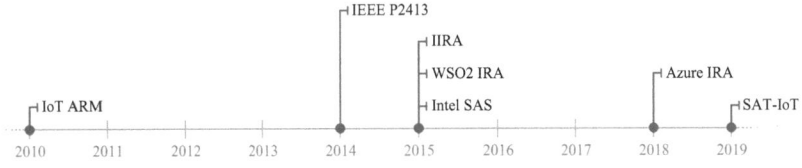

Fig. 1.2   Timeline of selected IoT Reference Architectures

1.4.1   Internet of Things Architectural Reference Model (IoT ARM)

The IoT-A project (IoT-A 2019) groups the specificities of IoT functionalities and defines the IoT Architectural Reference Model (IoT ARM) to support the usage, the development and the analysis of different IoT systems, from communication to service level.

According to Bauer et al. (2013), the main contributions of the IoT ARM are twofold: (a) the Reference Model itself, which contains a common understanding of the IoT domain and definitions of the main IoT entities and their basic relationships and interactions; and (b) the Reference Architecture *per se*, which provides views and perspectives to generate IoT architectures adapted to one's specific requirements. This way, the Reference Model and the Reference Architecture provide abstraction levels (models, views and perspectives) to derive concrete IoT solutions (i.e. IoT ARM compliant IoT architectures and systems) (Fig. 1.3).

The Reference Architecture is independent from a specific use-case or application and includes three views: (a) functional, (b) information, and (c) deployment and operation. The functional view describes the function components of a system; these include components' responsibilities, default functions, interfaces, and interactions. The architecture is composed of five longitudinal functionality groups (FGs), namely service organisation, IoT process management, virtual entity, IoT services, communication, and two transversal FGs, namely management and security.

The information view covers the information life cycle in the IoT system, providing an overview of the information structures and flows (i.e. how information is defined, structured, exchanged, processed, and stored), and the list of the components involved in the process.

Lastly, the deployment and operation view has an important role in the realisation of IoT systems as they are bringing together a number of devices, each of which has different resources and connection interfaces, which can be interconnected in numerous ways. The deployment and

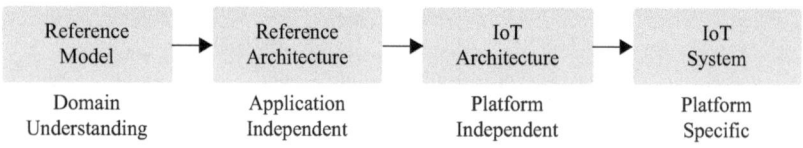

Fig. 1.3 Derivation from each IOT ARM step

operation view provides a set of guidelines for system design, covering different aspects of technologies, communication protocols, services, resources, and information storage.

According to Bauer et al. (2013), evolution and interoperability, availability and resilience, trust, security and privacy, and performance and scalability are the most important for perspectives for IoT systems.

Bauer et al. (2013) also present a reverse mapping to demonstrate how the concepts of the IoT ARM can be presented to existing architectures and to validate their proposal. One of the use cases was based on the use of RFID for tracing the towels before, during, and after the surgery to avoid towels being left on the patient's abdomen. This use case was also based on the use of a cloud infrastructure for data storing. Even though the authors argue that the IoT ARM mapping was successfully done, there is no way to say that it can be applied to any existing concrete architecture.

1.4.2   IEEE Standard for an Architectural Framework for the Internet of Things (P2413)

To avoid silos in domain-specific standards, P2413 is a unified architectural framework for IoT. As well as defining the framework, it includes descriptions of various IoT domains, definitions of IoT domain abstractions, and identification of commonalities between different IoT domains (energy, media, home, transport etc.). It provides a reference model that defines relationships among various IoT verticals and common architecture elements. In this way it has similar design principles to IoT ARM. The Reference Architecture covers the definition of basic architectural building blocks and their ability to be integrated into multi-tiered systems. The Reference Architecture also addresses how to document and mitigate architecture divergence. P2413 also includes a blueprint for data abstraction and addresses the need for trust through protection, security, privacy, and safety. Applying P2413, the architectural transparency of IoT systems can be improved to provide benchmarking, safety, and security assessments.

The P2413.1 is the Standard for a Reference Architecture for Smart City (RASC) (P2413.1 2019). The RASC provides an architectural design for the implementation of a smart city, enabling interaction and interoperability between domains and system components. The smart city applications may include water management, waste management, street lighting, smart parking, environmental monitoring, smart community, smart campus, smart buildings, e-health, e-government, etc. The RASC includes the Intelligent Operations Center (IoC) and IoT.

The P2413.2 is the Standard for a Reference Architecture for Power Distribution IoT (PDIoT) (P2413.2 2019). Following a similar idea of RASC, the PDIoT also provides an architectural design but for implementing power distribution systems, covering different domains, such as legacy grid systems, IoT and cloud computing. This standard defines a cloud based power distribution which supports microservices and migration from legacy systems to IoT based platforms.

1.4.3   Industrial Internet Reference Architecture (IIRA)

The term 'Industrial Internet' is largely attributed to General Electric (GE). In a joint report, Accenture and GE (2014, p. 7) define the industrial internet as an architecture that:

> [...] enables companies to use sensors, software, machine-to-machine learning and other technologies to gather and analyse data from physical objects or other large data streams—and then use those analyses to manage operations and in some cases to offer new, value-added services.

Today, the Industrial Internet has evolved in to the Industrial Internet of Things (IIoT). IIoT is defined Boyes et al. (2018, p. 3) as:

> A system comprising networked smart objects, cyber-physical assets, associated generic information technologies and optional cloud or edge computing platforms, which enable real-time, intelligent, and autonomous access, collection, analysis, communications, and exchange of process, product and/or service information, within the industrial environment, so as to optimise overall production value.

Somewhat like IoT ARM and P2413, the Industrial Internet Reference Architecture (IIRA) (Lin et al. 2019) is an architecture framework to develop interoperable IIoT systems for diverse applications across industrials verticals.

IIRA is composed of one frame and different representations (Fig. 1.4). According to (Lin et al. 2019), a frame is a collection of concepts represented by stakeholders (individual, team, organisation having interest in a

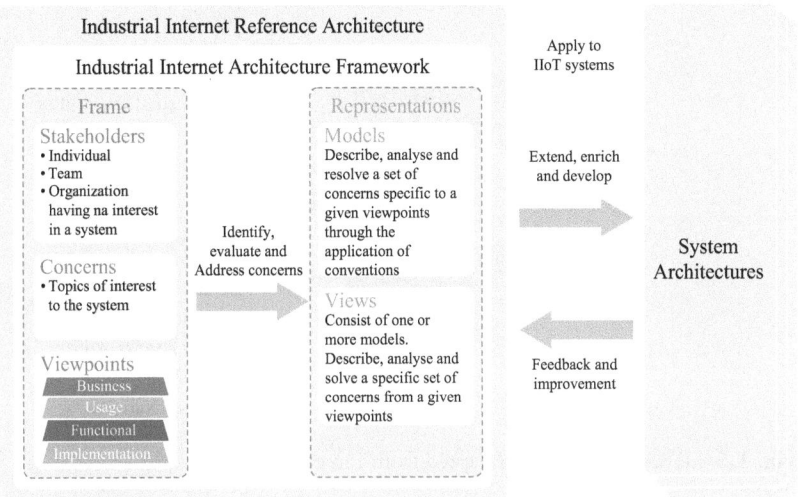

Fig. 1.4 Industrial internet Reference Architecture. (Adapted from Lin et al. 2019)

system), concerns (any topic of interest pertaining to the system), and viewpoints (conventions framing the description and analysis of specific system concerns). Representations are defined as views and models, which are collections of the results obtained through the application of the architecture frame to abstracted or concrete systems. These models and views are chosen for addressing a specific concern at an appropriate level of abstraction (Lin et al. 2019).

The IIRA identifies the main architectural concerns found in IIoT systems and classifies them into viewpoints related to their respective stakeholders. Viewpoints are critical components in the IIRA; there are four different viewpoints (Fig. 1.5). Firstly, the Business Viewpoint is responsible for inserting the vision, values, and objectives of business stakeholders in the commercial and regulatory context. Secondly, the Usage Viewpoint describes how an IIoT system realises its key capabilities, by providing the sequence of activities that coordinates the system components. Thirdly, the Functional Viewpoint relates the functional and structural capabilities of an IIoT system and its components. It is decomposed into five main functional domains: control domain, operation domain,

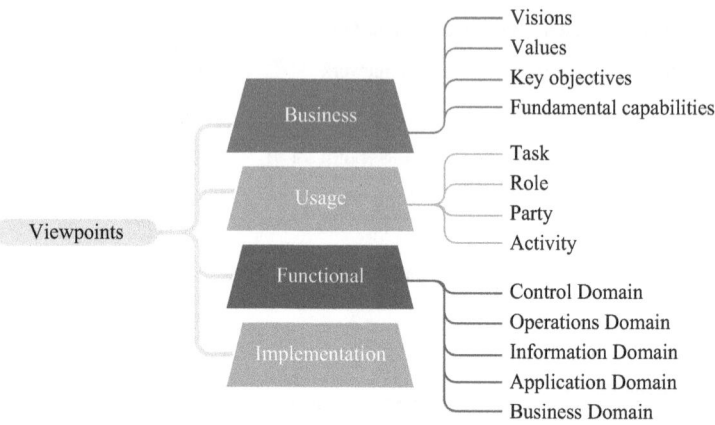

Fig. 1.5   IIRA viewpoints. (Adapted from Lin et al. 2019)

information domain, application domain and business domain. Finally, the Implementation Viewpoint provides (1) a description the general architecture of an IIoT system, (2) a technical description of its components, (3) an implementation map of the activities identified in the Usage Viewpoint; and (4) an implementation map for the key system characteristics (Lin et al. 2019).

By adopting IIRA, industries can integrate best practices into their processes, use a generic architecture and common framework and as a result reduce operation expenditure. It should be noted that IIRA provides architectural patterns for both cloud and edge computing.

1.4.4   WSO2 IoT Reference Architecture (WSO2 IRA)

WSO2 is a US-based open source integration vendor. The WSO2 IoT Reference Architecture (WSO2 IRA) is illustrated in Fig. 1.6 and supports IoT device monitoring, management, and interaction, covering the communication process between the IoT and the cloud (Fremantle 2015). The WSO2 IRA comprises five horizontal layers (client/external communication, event processing and analytics, aggregation layer, transports, and devices) and two cross-cutting layers (device management and identity and access management). Table 1.2 provides a brief definition of each layer.

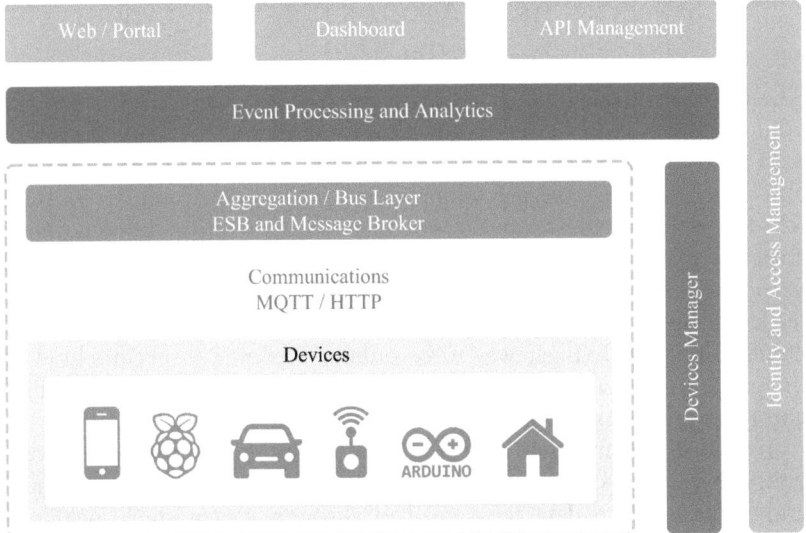

Fig. 1.6 WSO2 IoT Reference Architecture. (Adapted from Fremantle 2015)

Table 1.2 WSO2 IoT Reference Architecture layers

Layer	Description
Communication	Enables the devices to communicate outside of the device-oriented system web-based front-ends and portals, dashboards, and APIs.
Event processing and analytics	Takes the events from the bus and provides the ability to process and act upon these events.
Aggregation	Aggregates and brokers communications between devices, aggregates and combine communications from different devices and routes communications to a specific device, and bridges and transforms between different protocols.
Transport	Supports the connectivity of the devices.
Devices	IoT devices, they must have some communications that either indirectly or directly attaches to the Internet.
Device management	• Communicates with devices via various protocols and provides both individual and bulk control of devices. It also remotely manages software and applications deployed on the device. • Maintains the list of device identities and map these into owners. It must also work with the identity and access management layer to manage access controls over devices.
Access management	Provides identify and access management services.

1.5   Intel System Architecture Specifications (Intel SAS)

The purpose of the Intel System Architecture Specifications (SAS) is to connect any type of device to the cloud considering five key items: (1) C2T management, (2) real time analytics, (3) interoperability, (4) service and device discovery and provisioning, and (5) security (Intel 2015). Intel SAS has two distinct versions that co-exist in order to cover different infrastructure maturity levels: version 1.0 for connecting the unconnected and version 2.0 for smart and connected things. Version 1.0 specifies how legacy devices that were not originally designed to be connected to the cloud can use an IoT gateway to be online. Version 2.0 specifies how to integrate heterogeneous smart things focusing on security, manageability and real time data sharing between things and cloud (Fig. 1.7).

Intel SAS recommends a layered architecture that encompasses horizontal layers (users, runtime, and developers) and vertical layers (business and security). The data flow involves through eleven steps including analogue-to-digital conversion (ADC), gateways and reaching the cloud. Intel also recommends software components and interfaces to connect legacy devices with no connectivity functionality. The software components are located at endpoint devices and in the cloud. Basically, the cloud software components receive data collected by on-premise components and are responsible for analysis, storage, and service orchestration.

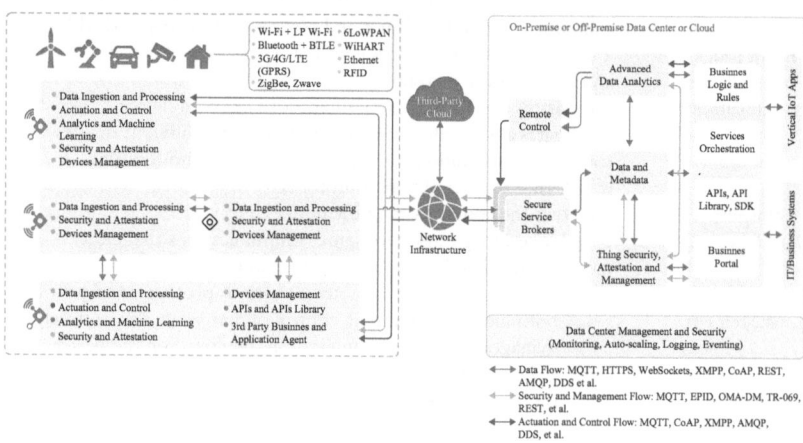

Fig. 1.7   Intel system architecture specifications. (Adapted from Intel 2015)

1.5.1 Azure IoT Reference Architecture (Azure IRA)

The Azure IoT Reference Architecture (Azure IRA) represented in Fig. 1.8 relies on Microsoft Azure platform to connect sensors to intelligent services at the cloud. The main goal of Azure IRA is to take actions on business insights that are generated through gathering data from IoT applications ('things') (Microsoft 2018). The reference document proposes a recommended IoT architecture, describing foundational concepts and principals, IoT subsystems details and solution design considerations. Azure IRA is focused on flexibility. As such, IoT solutions are cloud native and microservice-based. As deployable services are independent of each other, they suggest that it is better for scaling, updating individual IoT subsystems, and flexibility in the selection of technologies per IoT subsystem.

Figure 1.8 shows the recommended Azure IRA covering both hybrid cloud and edge solution integration. In orange, one can see the core IoT subsystems: IoT devices, cloud gateway (IoT Hub), stream processing, and user interface. The IoT device should be able to register with the cloud gateway, which is responsible for managing the devices. The stream processor consumes and stores the data, and integrates with the business process. For each subsystem, the Azure IRA recommends a specific technology based on Azure services. There is also a set of optional IoT

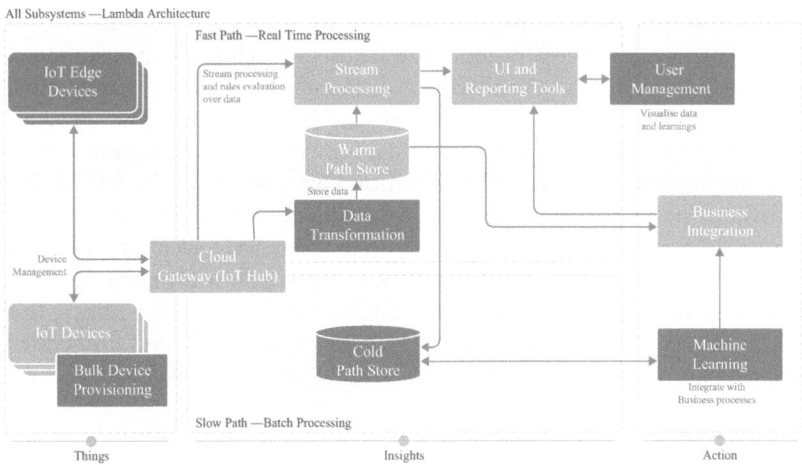

Fig. 1.8 Azure IoT Reference Architecture. (Adapted from Microsoft 2018)

subsystems (in blue): IoT edge devices, data transformation, machine learning, and user management. The edge devices are able to aggregate and/or transform and process the data on premise, while the data transformation (at the cloud) can manipulate and translate telemetry data. The machine learning subsystem allow the IoT system to learn from past data and act properly, such as firing alert to predictive maintenance. Finally, the user management subsystem provides functionality for users to manage the devices.

1.5.2   SAT-IoT

SAT IoT is a platform (Fig. 1.9) developed by Spanish company, SATEC, as part of the Horizon 2020 RECAP project.[1] Smart cities is a primary use case for SAT IoT. As such it needed an architecture that could (1) manage the smart city data network topology at run time, (2) use optimisation techniques that support processing aggregated data by geographical zones, and (3) monitor the IoT system and the optimisation process in run time (Peña and Fernández 2019).

Edge/cloud computing location transparency is a core feature of the platform allowing data to be shared between different zones (geographically and from the cloud to the edge), and thus to be processed at any of the edge nodes, mid nodes, or cloud nodes. This is realised by two of the entities in the SaT IoT architecture—the IoT Data Flow Dynamic Routing Entity and the Topology Management Entity. Together, they enable SAT IoT to manage the network topology at run time while also providing the necessary monitoring capabilities to understand the usage pattern and capacity limitations of the infrastructure. The IoT Data Flow Dynamic Routing Entity and the Topology Management Entity are augmented by the integration of the RECAP Application Optimiser in to SAT IoT, which derive the best possible placement of the data processing logic. Figure 1.9 shows the SAT-IoT architecture composed of Physical Layer, Smart Device Entity, IoT Data Flow Collector Entity, IoT Data Flow Dynamic Routing Entity, IoT Topology Management Entity, IoT Visualisation Entity, IoT Cloud Entity, Platform Access Entity, Security and Privacy, and Embedded IoT Applications.

---

[1] https://recap-project.eu/

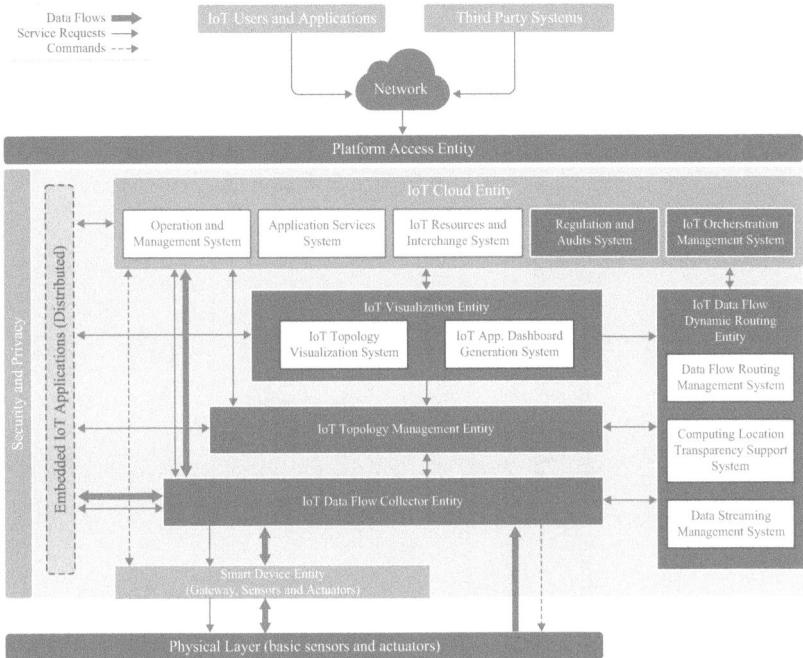

Fig. 1.9 The SAT-IoT Architecture. (Adapted from Peña and Fernández 2019)

1.5.3 Summary of Architectural Features

Table 1.3 summarises the key functional features addressed in each IoT Reference Architecture, that is interoperability, scalability, security and privacy, data management, analytics, data visualisation and user interface, and supported computing paradigms.

By system interoperability, we mean that the architecture should address connectivity, data management and automatic integration in a transparent way for the end user. Scalability refers to the architecture's ability to handle increases in the number of IoT devices and endpoints. Security and privacy capability ensures that the information be where it should be and prevents data leakage to unauthorised persons. Data management refers to both the management and exchange of data between architectural

Table 1.3 Summary of key features listed by different IoT Reference Architectures

Reference Architectures	Interoperability	Scalability	Security and privacy	Data management	Analytics	Data visualisation/ user interface	Supported computing paradigm
IoT ARM	X	X	X	X	–	–	IoT and cloud
IEEE P2413	X	X	X	X	–	–	IoT and cloud
IIRA	X	X	X	X	X	–	IoT, edge and cloud
WSO2 IRA	X	X	X	X	X	X	IoT and cloud
Intel SAS	X	X	X	X	X	X	IoT and cloud
Azure IRA	X	X	X	X	X	X	IoT edge and cloud
SAT-IoT	X	X	X	X	X	X	IoT, fog, edge and cloud

components. Analytics refers to the ability of the architecture to capture useful data from the deluge of data that travels on the network. Data visualisation and user interface is related to whether the architecture provides a human interface. Finally, computing paradigm refers to whether the architecture addresses support for new computing paradigms and specifically cloud, fog, edge, and dew computing.

Table 1.3 summarises the key features of different IoT Reference Architectures. It clearly emerges that only two functionalities are met by all Reference Architecture proposals—interoperability and security and privacy. Another common area of focus, unsurprisingly, is data management. Obviously, the primary value driver in the IoT is data and systems are required to manage the volume, velocity and variety of this data, not least where its stored and processes. The IEEE P2413 Reference Architecture presents less functionality; however this is due to the nature of such a standard. It is however the basis for a related smart cities standard (RASC).

When considering the IoT from a business, technical, or research perspective, each of these architecture features should be considered and addressed.

1.5.4   Conclusion

The chapter introduced two perspectives of the Internet of Things—a purely technical and a socio-technical perspective. The Internet of Things is not merely a technical phenomenon. It has the potential to transform how society operates and interacts. As such, it is critical to have a sufficiently general abstraction of the Internet of Things that facilitates sense making without getting in to a non-generalisable level of granularity. We present such an abstraction organised around five entities—social actors, things, data, networks, and events—and the processes that occur between them, all situated in time and space. We provided a brief overview of some of the key enabling technologies and new computing paradigms. Section 1.4 presented seven Reference Architectures for the Internet of Things and compared them across seven dimensions. This provides a further lens with which to consider the Internet of Things.

References

Accenture and General Electric. 2014. *Industrial Internet Insights Report for 2015.* Accessed April 2020. https://www.accenture.com/us-en/_acnmedia/Accenture/next-gen/reassembling-industry/pdf/Accenture-Industrial-Internet-Changing-Competitive-Landscape-Industries.pdf.

Bauer, Martin, Mathieu Boussard, Nicola Bui, Jourik De Loof, Carsten Magerkurth, Stefan Meissner, Andreas Nettsträter, Julinda Stefa, Matthias Thoma, and Joachim W. Walewski. 2013. IoT Reference Architecture. In *Enabling Things to Talk*, 163–211. Berlin, Heidelberg: Springer.

Boyes, Hugh, Bil Hallaq, Joe Cunningham, and Tim Watson. 2018. The Industrial Internet of Things (IIoT): An Analysis Framework. *Computers in Industry* 101: 1–12.

Breivold, Hongyu Pei. 2017. *A Survey and Analysis of Reference Architectures for the Internet-of-Things.* ICSEA 2017, 143.

Cavalcante, Everton, Marcelo P. Alves, Thais Batista, Flavia C. Delicato, and Paulo F. Pires. 2015. *An Analysis of Reference Architectures for the Internet of Things.* International Workshop on Exploring Component-based Techniques for Constructing Reference Architectures (CobRA).

Cisco. 2012. *The Internet of Everything: How More Relevant and Valuable Connections Will Change the World.* Cisco IBSG, 2012. Accessed December 2019. https://www.cisco.com/c/dam/global/en_my/assets/ciscoinnovate/pdfs/IoE.pdf.

Fremantle, Paul. 2015. *A Reference Architecture for the Internet of Things.* WSO2 White paper.

Gartner. 2017. Gartner Says 8.4 Billion Connected "Things" Will be in Use in 2017, Up 31 Percent from 2016. https://www.gartner.com/en/newsroom/press-releases/2017-02-07-gartner-says-8-billion-connected-things-will-be-in-use-in-2017-up-31-percent-from-2016.

Haller, S., S. Karnouskos, and C. Schroth. 2009. The Internet of Things in an Enterprise Context. In *Future Internet Symposium*, 14–28. Berlin, Heidelberg: Springer.

IEEE P2413. 2014. Standard for an Architectural Framework for the Internet of Things (IoT). *IEEE Standards Association*, 16 September. Accessed December 2019. http://grouper.ieee.org/groups/2413/Sept14_meeting_report-final.pdf.

Intel. 2015. The Intel IoT Platform Architecture Specification White Paper. Intel.

Iorga, Michaela, Larry Feldman, Robert Barton, Michael Martin, Nedim Goren, and Charif Mahmoudi. 2017. The NIST Definition of Fog Computing. *NIST Special Publication* (SP) 800-191 (Draft). National Institute of Standards and Technology.

IoT-A project. 2019. Accessed December 2019. https://cordis.europa.eu/project/id/257521.

IPv6 Addressing Architecture. 2003. https://tools.ietf.org/html/rfc3513.

Karygiannis, Tom T., Bernard Eydt, Greg Barber, Lynn Bunn, and T. Phillips. 2007. Guidelines for Securing Radio Frequency Identification (RFID) Systems: Recommendations of the National Institute of Standards and Technology. *NIST Special Publication*, 800-98.

Lin, Shi-Wan, Brandford Miller, Jacques Durand, Graham Bleakley, Amine Chigani, Robert Martin, Brett Murphy, and Mark Crawford. 2019. The Industrial Internet of Things Volume G1: Reference Architecture V1.90. *Industrial Internet Consortium*, June. Accessed December 2019. https://www.iiconsortium.org/pdf/IIRA-v1.9.pdf.

Lynn, Theodore, Philip Healy, Steven Kilroy, Graham Hunt, Lisa van der Werff, Shankar Venkatagiri, and John Morrison. 2015. *Towards a general research framework for social media research using big data*. In 2015 IEEE International Professional Communication Conference (IPCC), 1–8. IEEE.

Mell, Peter, and Tim Grance. 2011. The NIST Definition of Cloud Computing.

Microsoft. 2018. Microsoft Azure IoT Reference Architecture, Version 2.1. Accessed December 2019. http://download.microsoft.com/download/A/4/D/A4DAD253-BC21-41D3-B9D9-87D2AE6F0719/Microsoft_Azure_IoT_Reference_Architecture.pdf.

P2413.1. 2019. P2413.1 Standard for a Reference Architecture for Smart City (RASC). *IEEE Standard Association*. Accessed December 2019. https://standards.ieee.org/project/2413_1.html.

P2413.2. 2019. P2413.2 Standard for a Reference Architecture for Power Distribution IoT (PDIoT). *IEEE Standard Association*. Accessed December 2019. https://standards.ieee.org/project/2413_2.html.

Peña, Miguel Angel López, and Isabel Muñoz Fernández. 2019. *SAT-IoT: An Architectural Model for a High-Performance Fog/Edge/Cloud IoT Platform*. IEEE World Forum on Internet of Things (WF-IoT), 633–638. IEEE.

Shin, Donghee. 2014. A Socio-technical Framework for Internet-of-Things Design: A Human-Centered Design for the Internet of Things. *Telematics and Informatics* 31 (4): 519–531.

Stankovic, John A. 2008. When Sensor and Actuator Networks Cover the World. *ETRI Journal* 30 (5): 627–633.

Tarkoma, Sasu, and Artem Katasonov. 2011. *Internet of Things Strategic Research Agenda (IoT–SRA)*. Finnish Strategic Centre for Science, Technology, and Innovation: For Information and Communications (ICT) Services, Businesses, and Technologies, Finland.

Tsai, Shin-Yeh, Sok-Ian Sou, and Meng-Hsun Tsai. 2012. *Effect of Data Aggregation in M2M Networks*. In The 15th International Symposium on Wireless Personal Multimedia Communications, 95–99. IEEE.

Wang, Yingwei, and David Leblanc. 2016. *Integrating SaaS and SaaP with Dew Computing*. In 2016 IEEE International Conferences on Big Data and Cloud Computing (BDCloud), Social Computing and Networking (SocialCom), Sustainable Computing and Communications (SustainCom) (BDCloud-SocialCom-SustainCom), 590–594. IEEE.

Weyrich, Michael, and Christof Ebert. 2016. Reference Architectures for the Internet of Things. *IEEE Software* 33, no. 1: 112–116. Accessed December 2019. https://doi.org/10.1109/MS.2016.20.

Whitmore, Andrew, Anurag Agarwal, and Li Da Xu. 2015. The Internet of Things—A Survey of Topics and Trends. *Information Systems Frontiers* 17 (2): 261–274.

Zhao, Shanyang. 2003. Toward a Taxonomy of Copresence. *Presence: Teleoperators & Virtual Environments* 12 (5): 445–455.

CHAPTER 2

Next Generation Cloud Architectures

Konstantinos M. Giannoutakis,
Minas Spanopoulos-Karalexidis,
Christos K. Filelis Papadopoulos, and Dimitrios Tzovaras

Abstract This chapter discusses the evolution of cloud infrastructures and the impact of IoT. We discuss the need for heterogeneous resources integration in resource provisioning and the necessity to find the golden ratio between the cloud, fog and edge for optimal user experience. Complexity across the cloud-to-thing continuum is escalating rapidly. One way that cloud computing can adapt is to incorporate self-management and self-organization techniques and methods to preserve service availability and avoid user Quality of Service and Quality of Experience violations.

Keywords Cloud computing • Cloud architecture • Resource provisioning • Self-management • Self-organization

K. M. Giannoutakis (✉) • M. Spanopoulos-Karalexidis • D. Tzovaras
Information Technologies Institute, Centre for Research and Technology Hellas, Thessaloniki, Greece
e-mail: kgiannou@iti.gr; mspanopoulos@iti.gr; dimitrios.tzovaras@iti.gr

C. K. Filelis Papadopoulos
Department of Computer Science, University College Cork, Cork, Ireland
e-mail: christos.papadopoulos-filelis@cs.ucc.ie

2.1   Introduction

Cloud computing provides users with the potential to perform computing tasks utilizing resource physically distant to them. It offers virtually unlimited capabilities regarding storage, memory and processing units that augment IoT devices and components with limited computation capabilities due to form factors. However, this convergence of the cloud and the edge, provides significant issues, not least complexity at several orders of magnitude higher than the past. Forecast increases in the numbers of devices lead to a humongous escalation in generated data exacerbates this complexity. While Cloud-IoT cooperation seems perfect, managing the complexity across the a continuum can induce potential violations in Quality-of-Service (QoS) and Quality-of-Experience (QoE) user requirements. Thus, an evolution in conventional cloud infrastructure is necessary along with other infrastructure approaches such as fog and edge computing.

The remainder of this chapter is organized as follows. Section 2.2 provides a brief overview of the centralized cloud model and the significant impact of IoT on conventional cloud computing. Section 2.3 illustrates the changes in the centralized cloud paradigm. Next, decentralized cloud models are presented followed by a discussion of research opportunities and directions specifically focusing on the need for improved resource provisioning, support for resource heterogeneity, and self-management and self-organization patterns that cloud can adopt to address complexity. Concluding remarks are summarized in Sect. 2.6.

2.2   Centralized Cloud Computing Model

In the cloud computing era, the substance of a computer can no longer be limited to its physical "box" shape. Thus, a computer in the cloud has to introduce the capability of dynamically adjusting the required physical resources (i.e. processors, memory, storage, network bandwidth) based on any potential occasion. These resources can be distributed across physical servers and virtual machines creating a pool of available resources (Bhavani and Guruprasad 2014). This "cloud computer" offers increased performance levels, while dramatically reducing response time and operational costs.

2.2.1 Defining Cloud Computing

As defined in Chap. 1, cloud computing is

> *A model for enabling ubiquitous, convenient, on-demand network access to a shared pool of configurable computing resources (e.g., networks, servers, storage, applications, and services) that can be rapidly provisioned and released with minimal management effort or service provider interaction.* (Mell and Grance 2011, p. 2)

Conventional single provider infrastructures hosting cloud services on data centers offer a lot of benefits but also hide many challenges (Trilochan and Verma 2017). A large data center's energy consumption is high just to keep it operational and like any other centralized computing model, in case of a failure (single-point failure) the resulting issues would be adverse. Another issue is that required data may have to be transferred and stored to separate places, rather than the source, because data centers are often geographically distant from the application users. Thus, exchange of sensitive or personal data is considered critical for applications. For these reasons, alternate cloud infrastructure models which handle failures and use back up data centers have been introduced in recent years. Example models using cloud infrastructure rather than data centers are multi-cloud, micro cloud and cloudlet, ad hoc and heterogeneous clouds. The main cloud computing actors, characteristics and models are depicted in (Table 2.1).

A multi-cloud utilizes resources from multiple providers, thus making it possible to host large number of applications. Another form of multi-cloud is a federated cloud. This model is a federation of clouds from different cloud providers providing the potential to make applications portable, meaning that data from it or even the whole application can migrate from one cloud to another with the federation.

To incorporate resources located outside the cloud data centers. Micro cloud and cloudlet approaches offer a more decentralized computing infrastructure, as they are located closer to the edge of the network, thus minimizing transfer cost and communication latency as they are closer, sometimes a mere single hop, to the users and the generated data. Nevertheless, their computing power is much less than a conventional cloud infrastructure as they are composed of less powerful processors and are significantly smaller in size. An ad hoc cloud is an elastic infrastructure

Table 2.1   Cloud computing actors, essential characteristics, service models and deployment models. (Adapted from Liu et al. (2011) and Mell and Grance (2011))

Actors in cloud computing	
Consumer	A person or organization that maintains a business relationship with, and uses service from, cloud providers.
Cloud provider	A person, organization, or entity responsible for making a service available to interested parties. Cloud auditor A party that can conduct independent assessment of cloud services, information system operations, performance and security of the cloud implementation.
Cloud auditor	A party that can conduct independent assessment of cloud services, information system operations, performance and security of the cloud implementation.
Cloud broker	An entity that manages the use, performance and delivery of cloud services, and negotiates relationships between cloud providers and cloud consumers.
Cloud carrier	An intermediary that provides connectivity and transport of cloud services from cloud providers to cloud consumers.
Essential characteristics	
On-demand self-service	Consumers can unilaterally provision computing capabilities as needed automatically without requiring human interaction with the cloud provider.
Broad network access	Capabilities are available over the network and accessed through standard mechanisms that promote use by heterogeneous thin or thick client platforms and interfaces (e.g. devices).
Resource pooling	The provider's computing resources are pooled to serve multiple consumers using a multi-tenant model, with different physical and virtual resources dynamically assigned and reassigned according to consumer demand.
Rapid elasticity	Capabilities can be elastically provisioned and released, in some cases automatically, to scale rapidly outwards and inwards to meet demand. To the consumer, the capabilities available for provisioning often appear to be unlimited and can be appropriated in any quantity at any time.
Measured service	Cloud systems automatically control and optimise resource use by leveraging a metering capability at some level of abstraction appropriate to the type of service. Resource usage can be monitored, controlled, and reported, providing transparency to the service provider and the consumer.
Service models	
Software as a service	The capability provided to a consumer to use a provider's applications running on a cloud infrastructure and accessible by client interface.

(continued)

Table 2.1 (continued)

Platform as a service	The capability provided to a consumer to deploy onto the cloud infrastructure consumer-created or acquired applications created using development technologies provided by the provider.
Infrastructure as a service	The capability provided to a consumer to provision computing resources to deploy and run arbitrary software such as operating systems and applications.
Deployment models	
Private cloud	The cloud infrastructure is provisioned for exclusive use by a single organisation comprising multiple consumers. Ownership, management, and operation of the infrastructure may be done by one or more of the organisations in the community, by a third party, or a combination of both, and it may exist on or off premise.
Community cloud	The cloud infrastructure is provisioned for exclusive use by a specific community of consumers from organisations that have shared concerns. Ownership, management, and operation of the infrastructure may be done by one or more of the organisations in the community, by a third party, or a combination of both, and it may exist on or off premise.
Public cloud	The cloud infrastructure is provisioned for open use by the general public. It may be owned, managed, and operated by a business, academic, or government organisation, or some combination of them. It exists on the premises of the cloud provider.
Hybrid cloud	The cloud infrastructure is a composition of two or more distinct cloud infrastructures (private, community, or public) that remain unique entities, but are bound together by standardized or proprietary technology that enables data and application portability.

which deploys underutilized or spare resources of data centers, servers and any other mobile edge device in contribution to ad hoc computing (McGilvary et al. 2015). Finally, an heterogeneous cloud consists of different types of processing units in the infrastructure, mostly composed of accelerators such as Graphical Processing Units (GPUs), Intel Xeon Phis, Field-Programmable Gate Arrays (FPGAs) and others, offered in the form of VMs or containers. The problem is that, building a framework that utilizes and provides provisioning of these resources is still a challenging and difficult task (see e.g. the Horizon 2020 CloudLightning project (Filelis-Papadopoulos et al. 2018; Lynn et al. 2016; Xiong et al. 2017).

2.2.2   *The Impact of IoT on Conventional Cloud Architectures*

The evolution of the Internet of Things has had a significant impact on cloud computing generally and stretching the limitations of conventional cloud architecture. The number of connected devices is increasing exponentially with estimations of dozens of billions of "things" going live in coming years (Bittencourt et al. 2018; Gubbi et al. 2013).

The connected devices (things) are extremely heterogeneous in form and function. The trend to combine and integrate more and more sensors to each and every device escalates the complexity of resource management. As a consequence of connecting these sensors to the Internet, large volumes of data are being generated in unprecedented volumes, variety and velocity, the so-called Big Data (El-Seoud et al. 2017). This data is currently transferred and stored in the cloud in a centralized manner. Data transfer, especially in these volumes, is extremely expensive and retards computational performance. Thus, a more decentralized solution is required where data analysis could take place before transfer and storage. Emerging computing paradigms to support this decentralized or distributed cloud are discussed in detail in Sect. 2.4. Furthermore, most cloud infrastructures scale horizontally across multiple nodes in a data center or more thus making it necessary to develop cloud models that can scale vertically from low end processors to data center nodes.

The net result is that an enormous amount of data needs to be transmitted over the network, stored and/or processed by the receivers in an efficient way. The heterogeneity of the connected devices is immense and can be discrete in many levels, such as computing performance, storage and network requirements, communication protocols, energy consumption amongst others. This heterogeneity is adapted to many applications running on the IoT connected devices and as their numbers increase, so do their requirements accordingly, making it more complex and far more difficult to cope with the extensive needs that the computing system has to be able to accommodate.

2.3   Changes to Centralized Cloud Computing Model

More recently, application deployment in the cloud has been a challenge for the various providers in the network. To address this challenge, cloud architectures, infrastructure and deployment has evolved. This subsection provides a brief summary of a number of changes applied to these fields, including the evolution of service provisioning, the timeline of service models and the addition of new resources and workloads over time.

Monolithic Architectures are those architectures where the application is composed of a single program or platform, typically providing a user interface and data access through a database. Rationally, building and deploying an application of this model type is easy, especially when the application requirements are simple or the project is small. Nevertheless, the real challenge and the difficulties start to grow exponentially, when the application needs to scale up. Any adjustments, or any development and testing that needs to be made, initiates rebuilding of the whole code/project as it is a single unit or component. This could be extremely time consuming and potentially harmful, especially in large projects.

Service Oriented Architectures (SOA) are based on services to be provided in order to develop software product. Services are built to work in an orchestrated manner to modularize the system and provide a total service as a group. It is more challenging to divide the application in to multiple services, but it enables greater flexibility, extensibility and reusability of existing services for multiple use cases. The grouped services can communicate with each other by exchanging messages or events through APIs, which trigger the reaction of the corresponding services. The benefits of this model are application modularity, service reusability and enhanced security in the (re)building process and development of the application. A major disadvantage is the complexity in orchestrating all the services from a centralized component, especially when the project is complex and the components are huge.

Microservices introduced a solution for the gaps in the SOA approach. This approach divides applications in to more granular components by distributing them into small independent services. Each service implements and attends to separate business functions and capabilities to maintain independency from other services. They are the mainly deployed in an automated manner, through a container and communicating through RestAPIs, thus making the impact of programming language and data management techniques insignificant. This allows microservices to be easily deployed in the cloud, offering great reusability and minimal or no centralized management and orchestration. Essentially, microservices offer even more modularity than SOA and are more conducive in complex and large projects while, at the same time, providing independency in service development and maintenance, thus enhancing security in business products.

In addition to architecture modeling, different approaches have emerged regarding service modeling. In addition to conventional SaaS, IaaS, PaaS models, a new approach in virtualization is *bare metal or native virtualization* (Scarfone et al. 2011). Here, the hypervisor can run directly on the underlying hardware of the provider without a host operating system. Bare metal offers more security, providing that the hypervisor itself is well-secured. As the hypervisor is placed directly over the hardware, there is no host OS thus it cannot be harmed. This model is mostly used for servers in the cloud. Nevertheless, the hardware provided is limited so that a hypervisor does not consume the total available resources.

Hosted Virtualization is where the hypervisor runs over the host OS. The difference between hosted virtualization and bare metal is that the hypervisor is provided with more virtual resources but, on the other hand, the potential to harm the host OS is significantly increased.

Function as a Service (FaaS) is synonymous with serverless computing. Basically, FaaS enhances the microservices model development. During the development process, server operations are not taken into account, as services are hosted externally. Compared to conventional cloud models, where at least one server is utilized, FaaS triggers a server only when a function is conducted, executes the expected operations and then terminates. The major advantages of this model are increased scalability and independency of the applications and lower costs. As costs are based only on per used functionality, expenses from inactive resources are eliminated. A major disadvantage is the reduced transparency as FaaS is managed externally.

Composable architecture and Infrastructure is an approach used for provisioning both physical and virtual resources. It is an application-centric approach which greatly enhances operational performance with dynamic and flexible on-demand provision of resources. With the ability to manage a great variety of resources, it can easily scale up at an even greater extent than a traditional infrastructure. The flexibility composable infrastructure offers with the on-demand provision of resources, both hardware and software ones. Different resources can be provisioned independently (aggregated, disaggregated) and generally adjusted, based on their type (compute, storage, fabric) which is a major advantage of this approach.

Finally, until recently workloads and tasks on the cloud were largely loosely coupled. As such, the use of distributed memory was enough for the execution of any task, while the need for data rate and bandwidth was low. Thus, in a loosely coupled multiprocessor system, a simple message transfer system was enough to connect all the modules in the network. As the need for larger data rates and bandwidths are becoming more and more demanding especially with the addition of the heterogeneous resources in everyday tasks and applications, *tightly coupled workloads* in the essence of High Performance Computing (HPC) have been introduced in to cloud computing. These tasks are continuously more dependent on each other, utilize a common shared memory and the need for data rate and transfer is huge. A tightly coupled workload requires inter-process communication patterns that rely on high bandwidth with low latency in order to maintain optimal performance. This could lead to significantly reduced number of executed tasks, considering the demanding requirements.

2.4   DECENTRALIZED CLOUD COMPUTING MODEL

As the number of connected devices escalates rapidly, the generated data reaches unprecedented levels, and complexity increases as more and more sensors are integrated into user devices located at the edge of the network, conventional centralized data centers in the cloud can no longer provide an efficient and sustainable solution. It is necessary to provide resources and computational power closer to the edge users.

2.4.1   Fog Computing

Fog computing is a decentralized computing infrastructure which is used particularly as a complement to cloud computing. It leverages the compute resources at the edge network and brings the computational processing closer to the data source by offloading workload to edge nodes from cloud data centers. The network nodes near the edge providing these resources are called fog nodes. Overall, any device with computing, storage and network connectivity can constitute a fog node, for example switches and routers, industrial controllers, embedded servers and video surveillance cameras. A major benefit of fog computing is the reduction in application latency and as a result the improvement in QoS and QoE for users. Its first level usually lays a single hop away from the edge and is an extension of edge computing. Fog nodes can be organized in clusters, either vertically or horizontally depending on requirements, and can be either physical or virtual components that are tightly coupled with the end-devices. This introduces the need to be geographically aware in order to enhance performance and minimize end-user latency. Fog computing offers, similarly to conventional clouds, the aforementioned architectural service models (SaaS, PaaS and IaaS) and the deployment models (private, community, public, hybrid).

2.4.2   Mobile Edge Computing

Mobile Edge Computing, also referred to as just edge computing, takes place only on the edge of the network. Processing is executed closer to the data source and eliminates the costly data transfer to a remote data center or cloud (Ahmed and Ahmed 2016). This significantly improves user QoE as similar to fog computing, there is considerable network latency reduction and bandwidth consumption by the mobile subscribers.

2.4.3   Volunteer Computing

Volunteer Computing refers to a form of ad hoc cloud and cloudlets composed mostly of spare resources from users' computers or devices generally (Durrani and Shamsi 2014). The most common case where a VM is utilized in this manner is for social networks where users share their heterogeneous resources in the form of the aforementioned ad hoc cloud. The most reliable ones are rewarded and occasionally rewarded by payment for their contribution. Two significant challenges have to be overcome in order to fully benefit from volunteer computing. The first one is the difficulty to set up a reliable and functional virtualized environment considering the obstacles the heterogeneous resources and the ad hoc nature of this model. The second one is the privacy and security concerns users might have and need to be overcome in order to make them offer their spare resources and become volunteers.

2.4.4   Serverless Computing

Serverless Computing involves building, running and providing applications and services without taking into consideration the server side. "Serverless" does not mean that there is no server usage but rather the main focus on the application itself and the virtual resource provisioning in the hosting VM, rather than what happens on the physical infrastructure (Hellerstein et al. 2018). Serverless Computing is synonymous with FaaS and event-based programming as the execution of an application will be executed only when necessary and not all the time, thus meaning that an event can trigger the execution of a function or more than one function concurrently.

2.4.5 *Software-Defined Computing*

Software-defined computing separates the control plane from data plane and the utilized hardware in the network from the data control traffic components (Badotra and Singh 2017). This approach can also be adapted to other QoS metrics, rather than networking, such as storage and compute and resources located outside the data centers. It allows network managers to create a flexible, scalable pool of resources that are controlled by a software-driven cloud platform.

2.4.6 *Mist Computing*

Mist computing is a lightweight and elementary form of fog computing which resides even closer to the edge network and devices thus minimizing even more end-user latency (Iorga et al. 2018). It is comprised of dedicated nodes, with specialized capabilities but with less computational resources than fog. It is usually implemented as an extra layer of fog computing, closer or even on same layer with end-devices, but the existence of it is not considered mandatory.

2.5   RESEARCH DIRECTIONS AND OPPORTUNITIES

Cloud computing and associated technical evolutions is a solution to many IoT challenges but there are still obstacles that need to be overcome. The IoT-Cloud combination has to be able to provide accurate real-time processing and resource and service provisioning (Biswas and Giaffreda 2014). As already mentioned, these resources can be highly heterogeneous and require dynamic provisioning thus escalating complexity.

We identify five significant challenges for IoT-Cloud:

1. *Interoperability*—applications on a platform should be able to amalgamate services and infrastructure from another Cloud-IoT platform.

2. *Security and Privacy*—personal and sensitive user data are subjected to high risk while many users access public and ad hoc clouds. In some instances, personal data may have to be stored closer to the users/devices in order to facilitate computing and processing on the edge or fog layer. Furthermore, there is also the challenge to develop suitable and reliable encryption-decryption mechanisms and algorithms which could scale among distributed clouds and at the same time reduce energy consumption. This is discussed further in Chap. 6.

3. *Portability*—efficient migration of each application and service has to be supported from platform to platform and follow the users' traces and paths in the network.

4. *Reliability*—establishing real-time communication between objects and applications with high connectivity and accessibility

5. *Virtualization*—the potential to provision resources and provide access to heterogeneous resources and hardware such as GPUs, FPGAs, etc.

2.5.1   Resource Provisioning and Heterogeneous Resources

A central objective of cloud computing is to provide availability and share essential resources to each user to fulfill the QoS demands in respective SLAs (Parikh et al. 2017). Resource provisioning can be categorized into three main types:

1. *Static provisioning*—workloads and resource demands of an application are predefined or easily predictable.

2. *Dynamic provisioning*—is applied to applications that can adjust their demands during service, thus further resource allocation, or deallocation, may be needed or even migrate the application to another VM on-the-fly.

3. *Self-provisioning*—the user/application purchases specific resources from the cloud provider and can utilize them at will.

Efficient provisioning techniques target QoS metric improvement and reduce response times, SLA violations and power consumption. Furthermore, the cloud has to be operational even in case of a failure, that is being able to service user requests without making the failure noticeable to the outer world. This can be accomplished with optimal and novel remediation techniques in the cloud infrastructure (Ostberg et al. 2017). Similarly, resource provisioning in fog cells has to provide optimal utilization of fog resources and reduce offloading to cloud which dramatically increases costs (Skarlat et al. 2016). Additionally, data originated in the fog should be used in the particular fog and minimize data transfer between fog and cloud which significantly augments communication delays.

Evolving cloud infrastructures have introduced more heterogeneous resources, services and workloads while also augmenting complexity in resource provisioning. Heterogeneity can be discriminated in two types, machine heterogeneity and workload heterogeneity (Zhang et al. 2013). The addition of accelerators, such as GPUs and FPGAs, offer less powerful general purpose processors than a CPU but in distinctively greater numbers. These provide increased computational power and performance while reducing costs and improving energy efficiency. This is due to the potential of utilizing these many-core and multi-core systems for specific and specialized tasks like HPC. In this way, the overall computational and energy efficiency are noticeably increased. Workload analysis along with workload, machine and task heterogeneity can be characterized and classified (Zhang et al. 2014). A number of research studies have been undertaken in this domain including Kollenstart et al. (2018), Xiong et al. (2017), and Dai et al. (2015).

2.5.2 Self-management, Self-organization Approach

The ultimate goal of cloud computing as an ∗aaS (IaaS, PaaS, SaaS), is obviously to provide efficient services to users and meet their QoS requirements. Nevertheless, there are many occasions where failures occur and SLA violations occur. Additionally, massive changes in the cloud structure can also introduce significant performance challenges, for example flash crowd events, significant adjustment in the number of nodes. These challenges necessitate the development of adaptive patterns that can also address the increasing complexity of the cloud. Self-management techniques provide an approximate solution to the escalating complexity, as they tend to interact with both internal and external stimulus without any human intervention. They can be classified to four self-management aspects:

1. *Self-configuration*—the system manages the deployment of newly inserted nodes or disappearing ones by itself.
2. *Self-optimization*—a node or a network link hits its capacity limits, it has to be able to offload some of the tasks to another (optimal constituent) node/link,
3. *Self-protection*—the ability to protect itself against third party attacks, such as Distributed Denial-of-Service (DDoS), and
4. *Self-healing*—in the case of a failure, the active and executing applications have to be migrated and become available again elsewhere.

Generally, any cloud system has to guarantee its capacity to adapt and address the aforementioned challenges, that is continuous operation under any circumstances, load balancing, security, interoperability and energy efficiency. In order to achieve this, several patterns may be used based on service components and autonomic managers including self-organization and self-management (Xiong et al. 2017), P2P negotiation (Puviani et al. 2013) and centralized autonomic managers (Puviani and Frei 2013).

2.5.3   *Separation of Concerns*

Another major challenge derived from cloud computing utilization is the distinction between consumer-aware and provider-aware services respectively, along with a service interface establishment between them. This issue was first introduced by the CloudLightning project arising many considerable questions about services' origin and handling (Lynn et al. 2016; Xiong et al. 2017). Separation of Concerns the need to distinguish the exact services that users and providers should be concerned with respectively. Consumers should only be concerned with *what* they want to do and accomplish, and providers with *how* that could be done and provided to the user. A successful interface establishment between those two actors can lead to minimal direct consumer interaction with provider's infrastructure, thus allowing full control to the provider. Furthermore, this results in the assumption that various service implementations should already exist and the consumer himself does not have to be an expert and develop them. Additionally, physical resources provided by the infrastructure should not be consumer-aware and yet there may be several diverse implementations to meet specific service demands. These implementations can differ in hardware type and could be characterized by different price and performance attributes. Thus, consumers should be able to differentiate and discriminate between these implementations and choose the appropriate one that meets their service delivery attributes. In address such a difficult challenge, and to find the golden ratio between consumer and provider services, additional research is necessitated.

2.6 Conclusion

As a result of the Internet of Things and related technologies, cloud computing is experiencing a phase of rapid evolution. This chapter described how the Cloud-IoT convergence is moving cloud computing from a centralized model to a more distributed one and from a commoditized homogenous cloud to a specialized, heterogeneous one. New techniques and approaches are needed to exploit these new evolutions in cloud computing and to support the Internet of Things.

References

Ahmed, A., and E. Ahmed. 2016. *A Survey on Mobile Edge Computing.* 2016 10th International Conference on Intelligent Systems and Control (ISCO).

Badotra, S., and J. Singh. 2017. A Review Paper on Software Defined Networking. *International Journal of Advanced Research in Computer Science* 8 (2): 17.

Bhavani, B.H., and H.S. Guruprasad. 2014. Resource Provisioning Techniques in Cloud Computing Environment: A Survey. *International Journal of Research in Computer and Communication Technology* 3 (3): 395–401.

Biswas, A.R., and R. Giaffreda. 2014. *IoT and Cloud Convergence: Opportunities and Challenges.* 2014 IEEE World Forum on Internet of Things (WF-IoT). IEEE.

Bittencourt, L., R. Immich, R. Sakellariou, N. Fonseca, E. Madeira, M. Curado, L. Villas, L. DaSilva, C. Lee, and O. Rama. 2018. The Internet of Things, Fog and Cloud Continuum: Integration and Challenges. *Internet of Things* 3–4: 134–155.

Dai, W., H. Chen, and W. Wang. 2015. *RaHeC: A Mechanism of Resource Management for Heterogeneous Clouds.* 2015 IEEE 17th International Conference on High Performance Computing and Communications, 2015 IEEE 7th International Symposium on Cyberspace Safety and Security, 2015 IEEE 12th International Conference on Embedded Software and Systems, 40–45. IEEE.

Durrani, M.N., and J.A. Shamsi. 2014. Volunteer Computing: Requirements, Challenges, and Solutions. *Journal of Network and Computer Applications* 39: 369–380.

El-Seoud, S.A., H.F. El-Sofany, M. Abdelfattah, and R. Mohamed. 2017. Big Data and Cloud Computing: Trends and Challenges. *International Journal of Interactive Mobile Technologies (iJIM)* 11 (2): 34–52.

Filelis-Papadopoulos, Christos K., Konstantinos M. Giannoutakis, George A. Gravvanis, and Dimitrios Tzovaras. 2018. Large-scale Simulation of a Self-organizing Self-Management Cloud Computing Framework. *The Journal of Supercomputing* 74 (2): 530–550.

Gubbi, J., R. Buyya, S. Marusic, and M. Palaniswami. 2013. Internet of Things (IoT): A Vision, Architectural Elements, and Future Directions. *Future Generation Computer Systems* 29 (7): 1645–1660.

Hellerstein, J.M., J. Faleiro, J.E. Gonzalez, J. Schleier-Smith, V. Sreekanti, A. Tumanov, and C. Wu. 2018. Serverless Computing: One Step Forward, Two Steps Back. *arXiv preprint arXivL1812.03651*.

Iorga, M., L. Feldman, R. Barton, M.J. Martin, N. Goren, and C. Mahmoudi. 2018. *Fog Computing Conceptual Model*. U.S. Department of Commerce, Information Technology Laboratory. Gaithersburg: National Institute of Standards and Technology Special Publication 500-325.

Kollenstart, M., E. Harmsma, E. Langius, V. Andrikopoulos, and A. Lazovik. 2018. Adaptive Provisioning of Heterogeneous Cloud Resources for Big Data Processing. *Big Data and Cognitive Computing* 2 (3): 15.

Liu, F., J. Tong, J. Mao, R. Bohn, J. Messina, L. Badger, and D. Leaf. 2011. *NIST Cloud Computing Reference Architecture*. U.S. Department of Commerce, Information Technology Laboratory. Gaithersburg: National Institute of Standards and Technology Special Publication 500-292.

Lynn, T., H. Xiong, D. Dong, B. Momani, G. Gravvanis, C. Filelis-Papadopoulos, A. Elster, et al. 2016. Cloudlightning: A Framework for a Self-organising and Self-managing Heterogeneous Cloud. *CLOSER*: 1 (2): 333–338. https://dl.acm.org/doi/10.5220/0005921503330338

McGilvary, G.A., A. Barker, and M. Atkinson. 2015. *Ad hoc Cloud Computing*. 2015 IEEE 8th International Conference on Cloud Computing (CLOUD). New York: IEEE.

Mell, P., and Grance, T. 2011. The NIST definition of cloud computing, Recommendations of the National Istitute of Standards and Technology, NIST Special Publication 800-145.

Ostberg, P.-O., J. Byrne, P. Casari, P. Eardley, A.F. Anta, J. Forsman, J. Kennedy et al. 2017. *Reliable Capacity Provisioning for Distributed Cloud/Edge/Fog Computing Applications*. 2017 European Conference on Networks and Communications (EuCNC), 1–6. IEEE.

Parikh, S.M., N.M. Patel, and H.B. Prajapati. 2017. Resource Management in Cloud Computing: Classification and Taxonomy. *arXiv preprint arXiv:1703.00374*.

Puviani, M., and R. Frei. 2013. *Self-management for Cloud Computing*. Science and Information Conference, London.

Puviani, M., Cabri, G., and Zambonelli, F. (2013, July). *A Taxonomy of Architectural Patterns for Self-adaptive Systems*. In Proceedings of the International C* Conference on Computer Science and Software Engineering, 77–85.

Scarfone, K., M. Souppaya, and P. Hoffman. 2011. *Guide to Security for Full Virtualization Technologies*. U.S. Department of Commerce, Computer Security. Gaithersburg: National Institute of Standards and Technology Special Publication 800-125.

Skarlat, O., S. Schulte, M. Borkowski, and P. Leitner. 2016. *Resource Provisioning for IoT Services in the Fog*. 2016 IEEE 9th International Conference on Service-Oriented Computing and Applications (SOCA), 32–39. IEEE.

Trilochan, and A. Verma. 2017. Cloud Computing: Evolution and Challenges. *International Journal of Engineering Science and Computing* 7 (4): 10197–10200.

Xiong, H., D. Dong, C. Filelis-Papadopoulos, G.G. Castane, T. Lynn, D.C. Marinescu, and J.P. Morrison. 2017. *CloudLightning: A Self-organized Self-managed Heterogeneous Cloud*. Proceedings of the Federated Conference on Computer Science and Information Systems, 749–758. ACSIS.

Zhang, Q., M.F. Zhani, R. Boutaba, and J.L. Hellerstein. 2013. *HARMONY: Dynamic Heterogeneity-Aware Resource Provisioning in the Cloud*. 2013 IEEE 33rd International Conference on Distributed Computing Systems, 510–519. IEEE.

———. 2014. Dynamic Heterogeneity-Aware Resource Provisioning in the Cloud. *IEEE Transactions on Cloud Computing* 2 (1): 14–28.

CHAPTER 3

Flying to the Clouds: The Evolution of the 5G Radio Access Networks

Glauco E. Gonçalves, Guto L. Santos, Leylane Ferreira,
Élisson da S. Rocha, Lubnnia M. F. de Souza,
André L. C. Moreira, Judith Kelner, and Djamel Sadok

Abstract The number of connected devices and the amount of data traffic exchanged through mobile networks is expected to double in the near future. Long Term Evolution (LTE) and fifth generation (5G) technologies are evolving to support the increased volume, variety and velocity of data and new interfaces the Internet of Things demands. 5G goes beyond increasing data throughput, providing broader coverage and reliable

G. E. Gonçalves (✉)
Rural Federal University of Pernambuco, Recife, Brazil
e-mail: glauco.goncalves@ufrpe.br

D. Sadok • G. L. Santos • L. Ferreira • É. d. S. Rocha •
L. M. F. de Souza • A. L. C. Moreira • J. Kelner
Universidade Federal de Pernambuco, Recife, Brazil
e-mail: jamel@gprt.ufpe.br; guto.leoni@gprt.ufpe.br; leylane.silva@gprt.ufpe.br;
elisson.rocha@gprt.ufpe.br; lmfs@ecomp.poli.br; andre@gprt.ufpe.br;
jk@gprt.ufpe.br

© The Author(s) 2020
T. Lynn et al. (eds.), *The Cloud-to-Thing Continuum*, Palgrave
Studies in Digital Business & Enabling Technologies,
https://doi.org/10.1007/978-3-030-41110-7_3

41

ultra-low latency channels to support challenging future applications. However, this comes with a cost. As such, the architectural design of radio access network requires due consideration. This chapter explains why the radio access network is critical to 5G success and how novel trends on edge computing, network slicing and network virtualisation perform a critical role in optimising resources on emerging 5G infrastructures.

Keywords 5G • Network function virtualisation • Radio access networks • Cloud radio access networks

3.1   INTRODUCTION

The combination of widespread adoption of smartphones and the Internet of Things (IoT) presents telecommunications operators with significant challenges that legacy architectures were not designed to handle. An ever-increasing number of consumers use a plethora of bandwidth-intensive mobile applications, not least social media and video streaming, and device capabilities driven by affordable data plans. At the same time, the Internet of Things is driving data exchange; the number of smart end-points, for example, smart home and healthcare devices, will reach about 1.1 billion devices in 2022 (Cisco 2019). Globally, mobile devices and connections will grow to 12.3 billion by 2022 at a compound annual growth rate of 7.5% generating 77 exabytes (EB) of mobile traffic (Cisco 2019).

As discussed in Chap. 1, innovations such as Ipv6 and new paradigms in computing such as fog, edge and dew computing are enabling the IoT, however, LTE and 5G play a critical role in network connectivity. Furthermore, 5G, in particular will stimulate innovation and value through new applications and business models to support unprecedented connectivity and coverage. These applications and business models require increasingly heterogeneous and demanding service levels in terms of security, reliability, latency, throughput and so on (Li et al. 2018). In order to support these requirements, 5G technology evolves the 4G network through a new high frequency radio technology that provides greater data rates. Due to the smaller coverage of the high frequency radio technology, 5G needs more base stations to cover the same area than 4G, which in turn offers more resources to cope with the massive connectivity and low power demands of IoT devices. 5G technology can also 'slice' radio resources to offer more reliability, more bandwidth, or ultra-low latency according to the demand of the heterogeneous services coexisting within the 5G network (Popovski et al. 2018).

One of the main economic issues for operators of mobile infrastructure is that the average revenue per user (ARPU) is not growing as quickly as the traffic demand. As such, network operators are looking for mechanisms to sweat legacy infrastructure and reduce costs:

> *there has [...] been a need for cost-effective solutions that can help operators accommodate such huge amounts of mobile network traffic while keeping additional investment in the mobile infrastructure minimal.* (Taleb and Ksentini 2013, p. 12)

5G may be the answer. However, this may be a blessing in disguise. Firstly, while new business models and use cases may generate new value and revenue streams, it will also result in even greater heterogeneity, data, and QoS demands. Secondly, the cost of a 5G base station cost is estimated to be 4X of an equivalent Long-Term Evolution (LTE) base station and, due to the usage of higher frequencies, 5G is likely to need around 3 times more base stations to achieve the same coverage as 4G networks. Wisely et al. (2018) estimate that a 5G network with 100 times more capacity than a 4G network is 4 to 5 times more expensive than that 4G network. Finally, 5G's base station power consumption is estimated to reach 3X that of an LTE's. 5G uses massive multiple-input multiple-output (MIMO) antennas to perform beamforming and gain bandwidth. In contrast, LTE MIMO antennas usually use no more than 4 by 4 elements; 5G MIMO is expected to adopt 64 (at transmitter) by 64 (at receiver) antenna elements. It requires more power amplifiers and analogue-to-digital paths, and consequently increases power consumption to tens of kilowatts per base station. Clearly, the cost of deploying 5G is an important issue. Therefore, one solution is greater optimisation of the 5G radio access network (RAN) architecture in order to save resources. As a result, the telecommunication operators need to distribute their network infrastructure to the edge to cope with the growing number of mobile users and the related bandwidth-intensive mobile applications, minimising the communication path between users and services, and consequently decreasing the delay and alleviating pressure on core network operation. In this context, distributed cloud data centres, network virtualisation and slicing techniques (such as Software Defined Networking (SDN), Network Function Virtualisation (NFV), and Virtual Network Function (VNF)) perform critical roles in ensuring service availability, network enhancements and cost reduction. As Taleb et al. note

> *Along with recent and ongoing advances in cloud computing and their support of virtualised services, it has become promising to design flexible, scalable, and elastic 5G systems benefiting from advanced virtualisation techniques of cloud computing and exploiting recent advances relevant to network function virtualisation.* (Taleb et al. 2016, p. 84)

Understanding the components of distributed data centres (at both the infrastructure and application levels) and the relationship between them is very useful for analysing and optimising both infrastructure and resource placement for composing VNF chains.

In this chapter, we provide a summary of the evolution of 5G architectures and explain why RAN designs are critical to 5G success and consequently, the success of IoT. We describe how the components and their functionalities evolved over time to meet the user and application requirements. We also present how some key technologies, such as SDN and NFV, support the evolution of cellular networks. We conclude with current research challenges and opportunities in this area.

3.2   The Evolution of Radio Access Networks (RANs)

The section outlines the evolution of RANs from Distributed RANs to Cloud RANs, Heterogeneous Cloud RANs and Fog Computing RANs.

3.2.1   *Distributed Radio Access Networks*

Early generations of cellular systems used to have a baseband unit (BBU) and remote radio head (RRH) components physically integrated and located at the bottom of a Base Station (BS) connected to a Radio Frequency (RF) antenna at the top of the tower through heavy electrical cables. However, this architecture presented significant RF signal propagation loss in the electrical cable feed resulting in degraded signal transmission/reception power and quality (Liu 2017). As a result, telecommunications operators began to adopt a separated BBU and RRH architecture based on distributed Radio Access Network (D-RAN or just RAN).

In D-RAN, as shown in Fig. 3.1, each BS is composed of two co-located components: (1) a digital unit (DU) or BBU, and (2) a radio unit (RU) or RRH; these two components were connected through a Common Public Radio Interface (CPRI). The BBU is the component responsible for baseband processing, that is processing calls and forwarding traffic. The RRH is responsible for digital radio signal processing by transmitting, receiving and converting signals, as necessary. Each BS is connected to the core network through a backhaul.

In conventional D-RAN architectures, improving the operational capacity of a cell means to densify the network however this results in increased cost as additional BS' need be deployed and each BS has an

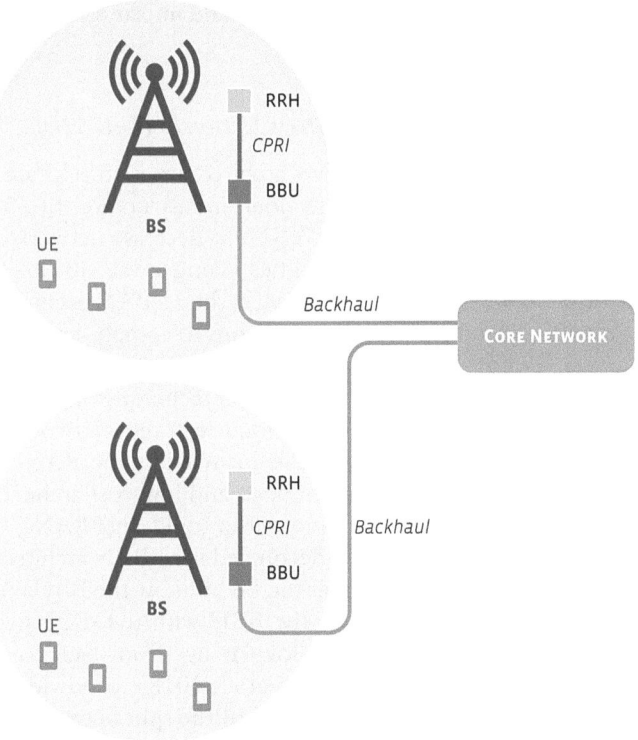

Fig. 3.1 A traditional D-RAN architecture

associated RRH and BBU. Additionally, in this scenario, the processing resources of a BBU cannot be shared among other RRHs.

Wang et al. (2017) propose an alternative means to improve the system by making use of technologies, such as coordinated multipoint (CoMP), to reduce the interference and increase the throughput. However, this solution has two main drawbacks: (1) it applies stringent delay constraints for control and signalling to guarantee on-time coordination between BS (NGMN Alliance 2015), and (2) it is not designed to deal with the processing capabilities of distributed BSs. As the volume of end users and complexity of the services offered by the operators has increased, new drawback in conventional D-RAN deployments emerged. For instance, average spectral efficiency gains of only 20% were observed in RAN deployments (Sun and Peng 2018). As a result, Cloud RANs (C-RANs) have emerged as a centralised solution, moving the BS functionalities to the cloud in order to optimise the resources and improve energy efficiency (Wu et al. 2015; Peng et al. 2015a, b).

3.2.2   Cloud Radio Access Networks (C-RANs)

The main design principle of C-RAN architecture is to relocate some of the cellular network functions to the cloud infrastructure. In 2010, IBM proposed a wireless network cloud (WNC) to decrease network costs and obtain more flexible network capabilities (Peng et al. 2011). In 2011, China Mobile Research Institute launched the C-RAN architecture and ZTE Corporation proposed network solutions to comply with the C-RAN requirements. Following this lead, many telecom operators started to develop new solutions based on virtualisation techniques in order to guarantee flexibility and take advantage of cloud features. Network operators understand that the main cost of 5G is incurred at the RAN, therefore they decided to invest in new types of open and low-cost architectures.

To address the main limitations present in traditional RANs, the RRH and BBU functions were physically decoupled in C-RAN architectures. As shown in Fig. 3.2, the RRH is kept at the BS but now the BBU is migrated to cloud infrastructure. To connect the BBU with the respective RRHs, there is a need for a high-speed and low-latency front-haul communication channel (Hossain et al. 2019; Ren et al. 2018); bandwidth requirements for these links depends on the level of the split between BBU pool located in the cloud and RRH. As presented in (Peng et al. 2015a, b), there are three different functional split of C-RAN architectures: (1) fully

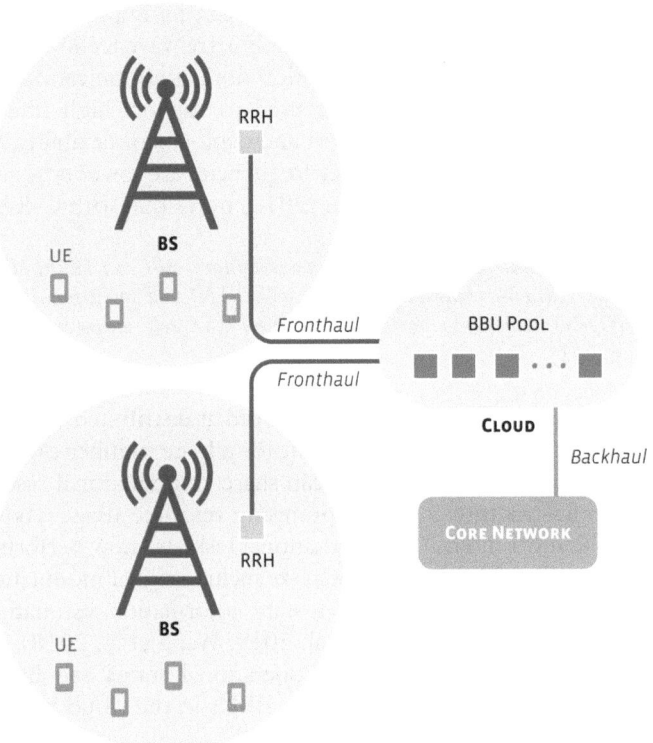

Fig. 3.2 A general architecture of a C-RAN cellular network

centralised, (2) partially centralised and (3) hybrid. In the fully centralised, all processing and management functions of base stations are performed by the BBU pool at the cloud. This way, basically all data need to be transferred from RRH to the cloud, requiring a high bandwidth. In the partially centralised configuration, the RRH performs the functions related to RF, such as signal processing; and the other management functions are performed in the cloud by the BBU pool. This option reduces the bandwidth requirements between the RRH and the cloud. However, the interaction between processing and managing functions can be complex, making the separation difficult. In this case, the third type of split, the hybrid, moves some types of processing functions to the cloud and assigns them to a new separated process. This option facilitates the resource management and reduces the energy consumption on the cloud side.

The front-haul communication channel can be implemented using a wide variety of technologies including millimetre wave technologies, standard wireless communication, and optical fibre communication (Hossain et al. 2019). While fibre optics are used to support high transmission capacity, these are constrained by cost and deployment flexibility. Wireless technologies with 5–40 GHz carrier frequencies are lower cost and more flexible in terms of deployment. Liu (2017) notes that with C-RANS:

> ...*the conventional complicated and power-hungry cells can be simplified to RRH only, reducing capital expenditures (CAPEX) and operational expenditures (OpEx) related to power consumption and cell maintenance.* (Liu 2017, p. 221)

In this way, several RRHs can be deployed at distributed BS to provide seamless coverage and high throughput for a large number of users (Pan et al. 2018), while a pool of BBUs can share computational resources in the cloud infrastructure thereby optimising resource usage. As such, in C-RANS, the most intensive computational tasks are now performance in BBUs allocated in the cloud. These tasks include signal modulation, precoding matrix calculation, channel state information estimation, and Fourier transformation (Hossain et al. 2019; Wang et al. 2018). In addition, the monitoring of the RRHs operational status can be used to dynamically adapt the number of active BBUs in the cloud reducing the energy and operational cost (Pan et al. 2018).

As highlighted in Hossain et al. (2019), there are several advantages in adopting C-RAN architectures. In a traditional RAN architecture, the deployment and the commissioning of a new BS is very expensive and time-consuming. In contrast, in C-RAN systems, the deployment of an equivalent infrastructure is relatively easier since only a new RRH need be installed and associated BBU services deployed in the cloud. With this, it is possible to cover new areas or split the cell in order to improve its capacity. Suryaprakash et al. (2015) suggests that the adoption of C-RAN can reduce CAPEX by approx. 15%. Furthermore, it is possible to improve energy efficiency. As all BBUs are allocated in the cloud, the telecommunications operator is able to monitor the BBUs operation and apply strategies to dynamically change their mode (low power sleep mode or shut down) to save energy saving energy (Wu et al. 2015).

There are some drawbacks in adopting C-RANs, not least security. C-RAN architectures may suffer the same problems of traditional networks, such as primary user emulation attack and spectrum sensing data falsification (Tian et al. 2017). In addition, if all BBUs run in the cloud, any problem in the cloud infrastructure can compromise the whole service operation. Peng et al. (2016a) note that centralised signal processing in the cloud can introduce the risk of higher latency. The constrained capacity of front-haul links is also a problem. This results in a significant negative impact on both energy efficiency and spectral efficiency (Sun and Peng 2018). Two proposed innovation to address these issues are heterogeneous C-RAN (H-CRAN) and fog RAN (F-RAN).

3.2.3 Heterogeneous Cloud Radio Access Networks

Heterogeneous CRAN (H-CRAN) is an architecture that takes advantage of two approaches: CRAN and Heterogeneous Networks (HetNets). HetNets are composed of a set of small cells that transmit signals with low power within a traditional macro cell network (Anpalagan et al. 2015). Hetnets allows short radio transmission distance resulting in reduced cost and promotes capacity enhancement (Yang et al. 2015).

Small cells can be classified as microcells, picocells, or femtocells. These types of cells are differentiated by output power, cell radius, number of users, and Distributed Antenna Systems (DAS) integration (see Table 3.1). DAS is a distributed version of a MIMO system that aims to provide spatial diversity to avoid path loss and shadowing. Consequently, the signal reception quality and the physical layer security at receivers are improved (Wang et al. 2016). H-CRANs, as well as HetNets, present different types of small cells in their architecture, which are spread along a macro cell coverage area (Marotta et al. 2017).

HetNets have two important types of nodes: high power nodes (HPNs) and low power nodes (LPNs). HPNs, such as macro cell base stations (MBS), are in charge of wide network coverage. LPNs, such as small cell

Table 3.1 Cell specification. (Adapted from Mishra (2018))

	Femto cells	Pico cells	Micro cells	Macro cells
Output power	1–250 mW	250 mW–1 W	1–10 W	10–50+ W
cell radius	10–100 m	100–200 m	0.2–2 km	8–30 km
Users	1–30	30–100	100–2000	2000+
DAS integration	No	Yes	Yes	Yes

base stations (SBS), are low powered nodes densely deployed, offering high data rates in hot spots and seamless mobility, referring to the SBSs (Sun and Peng 2018). To manage complexity and for efficiency and cost-effectiveness, HetNets support self-organisation, allowing cooperation between the base stations to optimally coordinate their resources.

The use of HetNet HPNs brings advantages to C-RAN architectures in terms of backward compatibility and seamless coverage in cellular networks, since in a C-RAN architecture, RRHs focus on high capacity instead of coverage. Furthermore, HPNs enable convergence of multiple heterogeneous radio networks and control signaling in the network (Alimi et al. 2017). In H-CRAN architectures, RRHs assume the role of LPNs by performing simple functions (such as radio frequency management and simple symbol processing). The BBU is responsible for coordination between HPNs and RRHs to mitigate inter-tier interference. The BBU pool is also responsible for important upper layer functions (Sun and Peng 2018; Ali et al. 2017).

In H-CRANs, the control and data plane are decoupled. Data rate is the responsibility of an RRH (LPN) while control plane functionality is allocated to HPNs (Zhang and Wang 2016; Ali et al. 2017). Figure 3.3 presents an H-CRAN architecture and its elements. The HPN located in the macro cell communicates with SBS' through the control plane. RRHs located in small cells communicate by front-haul with the BBU pool through the data plane. In this architecture, the communication from the HPN to the cloud, from the cloud to the core network, and from the core network to the HPN are done by the back-haul channel.

3.2.4   Fog Computing Radio Access Networks

Fog Computing Radio Access Network (F-RAN) exploits the edge and storage capabilities of fog computing to address the front-haul constraints of previous architectures C-RANs and H-CRANs. The C-RAN and H-CRAN architectures centralise their software process at the cloud resulting in a heavy load on the front-haul link. To mitigate this problem, Peng et al. (2016b) proposed the F-RAN architecture based on the H-CRAN architecture with the addition of two components: (1) a fog computing-based access point (F-AP), RRH equipment with caching, cooperative signal processing, and radio resource management (RRM); and (2) fog user equipment (F-UE), a smart user terminal that also contains caching, cooperative signal processing, and RRM. With both

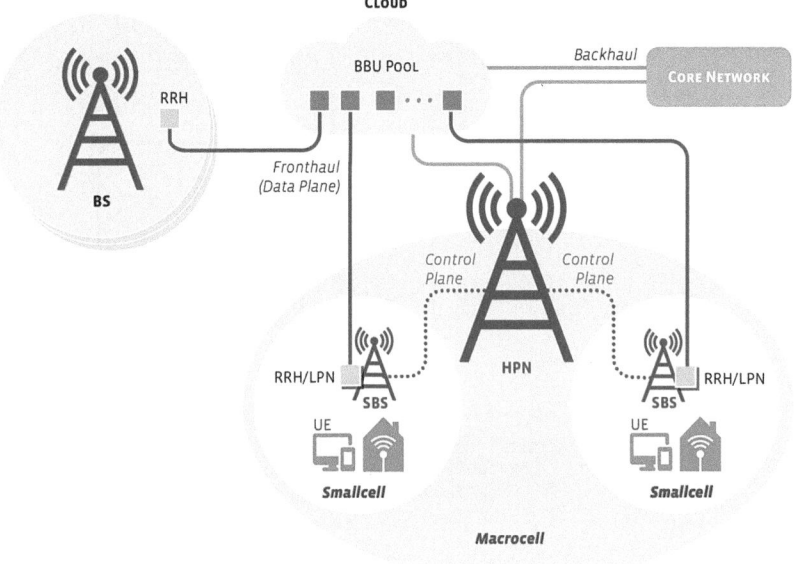

Fig. 3.3 An H-CRAN architecture

components, the proposed architecture (Fig. 3.4) receives local traffic from the F-APs or F-UEs, preventing traffic overloads from the front-haul to the cloud BBU.

The F-UEs can communicate to each other through an adaptive technique device-to device (D2D) or using the F-UE based relay mode. For instance, the F-UE can exchange data directly with another F-UE using the D2D technology (Peng et al. 2016b). Meanwhile, the relay mode uses an F-UE as intermediary communication to other F-UEs. As mentioned earlier, the F-APs are RRH equipment that store a content cache and are used to forward and process incoming data. Because the F-APs and F-UEs contain caching, the control plane and part of the data plane can be transferred to them. As such, some requests will be processed locally addressing front-haul limitations (Peng et al. 2016b).

Although the F-RAN aims to minimise the disadvantages of C-RAN and H-CRAN, some questions about the new architecture are still open, such as caching, SDN and NFV. Caching on F-AP and F-UE devices requires intelligent resource allocation strategies to be efficient and thus

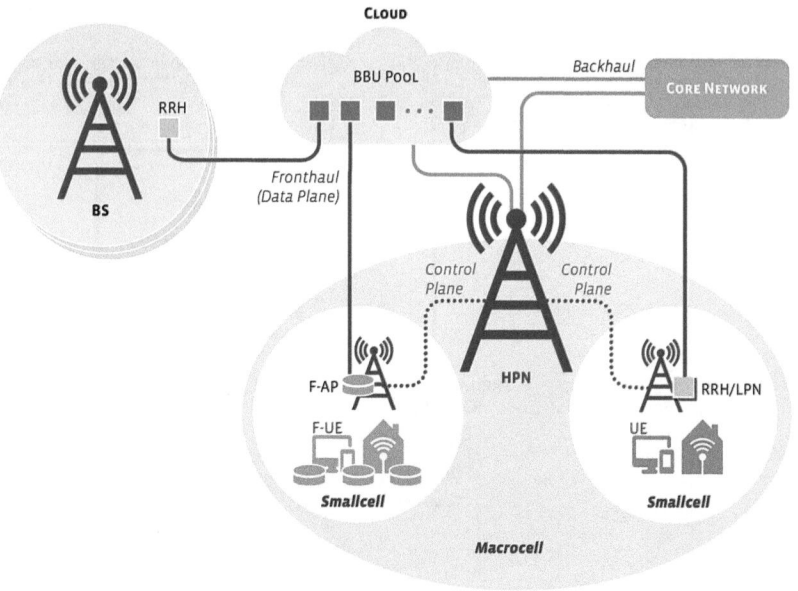

Fig. 3.4   An F-RAN architecture

alleviate front-haul overhead. Device caching is limited and can save little data locally. Thus, if both resource allocation and caching are not efficient, using F-RAN will not make sense and will not help with front-haul relief (Peng et al. 2016b). SDN, originally designed to be applied in wired networks, has been adapted for use in F-RANs. However, its structure is based on a centralised operation, while the F-RAN is based on a distributed one. As such, SDN needs to be adapted to this new context. In the same way, virtualising the SDN controller in F-RAN architectures remains a challenge (Guizani and Hamdi 2017).

3.3   NETWORK FUNCTION VIRTUALISATION AND 5G NETWORKS

5G networks deliver six benefits: high capacity, high data rate, low end-to-end latency, reduced costs, improvement of energy efficiency, and massive device connectivity (Zhang et al. 2015). Consequently, it needs ultra-densified networks, device-centric architecture, and specialised hardware.

There is also a need to coexist with legacy infrastructures, e.g. 2G, 3G, and 4G technologies, which increases management cost and complexity. A solution to address these factors is to implement the 5G network functions as software components using NFV (Alimi et al. 2017).

The initial purpose of NFV was to reduce CapEx and OpEx using virtualisation technology and to allow network operators sweat legacy infrastructure. NFV implements, through virtualisation technologies and leveraging standard servers, network functions in software, instead of running them on purpose-built hardware (Gomes et al. 2015). SDN enables the network operator to manage network functions through the abstract lower-level functionality, separating the control plane and data plane. At the same time, NFV is a technology that enables flexible and fast deployment of network functions in commodity devices instead of dedicated purpose-built hardware (Zeng et al. 2017). The combination of NFV and SDN brings several advantages for the network operator such as energy efficiency, network programmability (Miozzo et al. 2018; De Souza et al. 2018), network slicing (Ordonez-Lucena et al. 2017; Chartsias et al. 2017; Zhang et al. 2015; ETSI 2013; Zhou et al. 2016; Schiller et al. 2015), and dynamic bandwidth adjustment to reduce the delay (Zhang et al. 2015; Jahan 2015). For example, it is possible to identify the optimal resources to meet a specific demand and allocate them into the network using SDN/NFV (De Souza et al. 2018). SDN/NFV is increasingly being adopted by network operators not only for reduced CapEx and OpEx but also because it offers new service and revenue generation opportunities from legacy infrastructure by reducing the maturation cycle, deploying services faster (reduced time to market), and targeting particular customer segments or geographic markets with specific software and software configurations (Lynn et al. 2018).

In an effort to improve C-RANs, NFV has been used to virtualise the RAN architecture (ETSI 2013; Peng et al. 2015a, b; Rost et al. 2014; Dawson et al. 2014; Peña et al. 2019). ETSI outline a virtualised RAN use case in a C-RAN architecture where the BBU functions can be executed in a Network Function Virtualisation Infrastructure (NFVI) environment, such as a data centre. Peng et al. (2015a) used a H-CRAN solution based on NFV that included virtualised radio and computing resources for both intra and inter RAN technologies. Rost et al. (2014) proposed RAN-as-a-Service (RANaaS) to ensure flexible a functional split between a centralised cloud (e.g. C-RAN) and a distributed operation (in conventional mobile networks). They sought to take advantage of the flexibility of virtualised RAN functions, while delay-stringent functions remained at the BS' with the less stringent ones deployed centrally in the cloud. Dawson et al. (2014) proposed a virtual network architecture for Cloud-RAN base stations that presents the core network with an abstracted view of the physical network. Abdelwahab et al. (2016) explored the potential of NFV for enhancing the functional, architectural, and commercial feasibility of 5G RANs including increased automation, operational agility, and reduced CapEx.

The RECAP project developed the next generation of cloud, edge and fog computing resource management, that supports complex applications and networks, and make use of network and service function virtualisation to handle heterogeneous underlying architectures and dynamic resource provisioning. Representative uses cases were proposed to demonstrate the challenges and one of the use cases is owned by TIETO, the largest IT service company in the Nordics. TIETO provides new solutions leveraging on the possibilities enabled by 4G and beyond mobile technologies in conjunction with cloud and fog computing. Through the RECAP project, TIETO evaluated 5G technologies by simulating network characteristics and QoS requirements, focused on improving reliability and reducing network latency. The RECAP solution for TIETO relies on SDN and VNF to dynamically provide resources (application placement and infrastructure optimisation) considering the QoS and QoE requirements (Peña et al. 2019).

3.4   CHALLENGES AND FUTURE DIRECTIONS

C-RAN is currently established as an alternative to the distributed cellular RAN. It centralises mobile network functions and is shown to consistently reduce capital and operational expenditures of such networks. Despite this, there is currently a number of opportunities for architectural improvements.

In order to meet the requirements of 5G deployments, C-RAN technology must evolve to reduce the costs of high-speed front-haul networks. CPRI-based front-haul demands high data rates (typically 10 to 24 Gbps per RRH) and small latency ($100 \ \mu s$ to $400 \ \mu s$) due to the nature of the I/Q data (Gomes et al. 2015). As such, options like Ethernet-based links appear as cost effective alternatives to replace CPRI as they are based on low-cost equipment and it brings statistical multiplexing capabilities to the front-haul. Despite offering high data rates, Ethernet presents delay and synchronisation issues that remain as barriers to further adoption.

Greater energy efficiency is critical for future 5G mobile networks. The deployment of small 5G cells and heterogeneous networks will increase network energy demands. Harvesting ambient energy (through solar and wind power technology) are needed to make such deployments economically feasible and environmentally sustainable thus reducing energy consumption. At the same time, strategies to conserve energy at BBUs and RRHs (sleep mode) will be more and more employed (Hossain et al. 2019).

Even though the advantages of H-CRAN are well-documented, there are some open challenges in terms of operability. Front-haul and backhaul links may suffer additional burden due to the increasingly massive volumes of data received by the BBU pool (Zhang et al. 2017). The high density of base stations also may result in issues in H-CRAN architecture, such as inefficient resource usage, signal interference and degraded throughput in cases where distant cells are located at the cloud edge (Tran et al. 2017).

Supporting a massive amount of device-to-device communications brings several challenges that must be overcome in order to make 5G radio access the main infrastructure for the IoT. First, new IoT services and applications will change the traffic matrix at the RAN, as there will be an increase in connections between devices at the edge and between these devices and the distributed applications hosted close to the BBU. Such a traffic matrix will lead the front-haul to change uplink and downlink requirements. Second, the sheer mass of new IoT devices will bring new mobility management issues due to the increase in handoff and location operations. This will, in turn, be impacted by the centralised nature of C-RANs that can impose additional latency to perform these operations. To manage the complexity inherent in such a massive volume of heterogeneous and geographically distributed end-points, self-organisation presents itself as a solution and an avenue for further research (Hossain et al. 2019).

In terms of security, C-RAN technology is subject to threats from cloud systems and cellular systems (Hossain et al. 2019). Research in this area needs to employ security frameworks and techniques from both worlds (cloud and cellular) to promote new solutions for maintaining user privacy, trust among devices in HetNets, and among devices from different operators. The security challenges also extend to physical security. Wireless communications, by their nature, are susceptible to eavesdropping, and standard solutions based on encryption often impose infeasible or unacceptable computing and communication overheads. This way, development of strategies to exploit the physical characteristics of the radio channel for security is an active research field (Peng et al. 2016a)

3.5   Conclusion

Radio access networks, and 5G technologies in particular, provide the network connectivity to enable the Internet of Things. In this chapter, a survey is presented on the evolution and improvements of radio access networks for 5G cellular networks (D-RAN, C-RAN, H-CRAN, and F-RAN) by presenting their infrastructure details, advantages, and limitations. A selection of key emerging technologies, such as SDN and NFV, and their benefits are also discussed. 5G deployments, energy efficiency, massive device-to-device communications, and security in RAN-based architectures all present potentially fruitful and necessary avenues for research as the adoption of the Internet of Things accelerates. We believe that this survey serves as a guideline for future research in 5G networks, as well as a motivator to think about on the next generation 5G RAN architectures for the Internet of Things.

References

Abdelwahab, S., B. Hamdaoui, M. Guizani, and T. Znati. 2016. Network Function Virtualization in 5G. *IEEE Communications Magazine* 54: 84–91.

Ali, M., Q. Rabbani, M. Naeem, S. Qaisar, and F. Qamar. 2017. Joint User Association, Power Allocation, and Throughput Maximization in 5G H-CRAN Networks. *IEEE Transactions on Vehicular Technology* 66 (10): 9254–9262.

Alimi, Isiaka A., P.M. Paulo, and L.T. António. 2017. Outage Probability of Multiuser Mixed RF/FSO Relay Schemes for Heterogeneous Cloud Radio Access Networks (H-CRANs). *Wireless Personal Communications* 95 (1): 27–41.

Alliance, N.G.M.N. 2015. Ran Evolution Project Comp Evaluation And Enhancement. Accessed 10 March 2019. https://pdfs.semanticscholar.org/8e35/148 ab1530342ca7c3659bfbfbe5c1e454945.pdf?_ga=2.25742072. 478265301.1575550962-2016605258.1575550962

Anpalagan, A., B. Mehdi, and V. Rath. 2015. *Design and Deployment of Small Cell Networks*. Cambridge University Press.

Chartsias, P. K., A. Amiras, I. Plevrakis, I. Samaras, K. Katsaros, D. Kritharidis, E. Trouva, I. Angelopoulos, A. Kourtis, M. S. Siddiqui, A. Viñes, and E. Escalona. 2017. *SDN/NFV-based End to End Network Slicing for 5G Multi-tenant Networks*. Paper presented at European Conference on Networks and Communications (EuCNC). IEEE. https://ieeexplore.ieee.org/document/7980670.

Cisco. 2019. Cisco Visual Networking Index: Global Mobile Data Traffic Forecast Update, 2017–2022. https://www.cisco.com/c/en/us/solutions/collateral/service-provider/visual-networking-index-vni/white-paper-c11-738429.pdf

Dawson, A.W., M.K. Marina, and F.J. Garcia. 2014. *On the Benefits of RAN Virtualisation in C-RAN based Mobile Networks.* Paper presented at Third European Workshop on Software Defined Networks, London, 103–108.

De Souza, Phelipe A., Abdallah S. Abdallah, Elivelton F. Bueno, and Kleber V. Cardoso. 2018. *Virtualized Radio Access Networks: Centralization, Allocation, and Positioning of Resources.* IEEE International Conference on Communications Workshops (ICC Workshops), May.

Gomes, N.J., P. Chanclou, P. Turnbull, A. Magee, and V. Jungnickel. 2015. Fronthaul/Evolution: From CPRI to Ethernet. *Optical Fiber Technology* 26: 50–58.

Guizani, Z., and N. Hamdi. 2017. CRAN, H-CRAN, and F-RAN for 5G Systems: Key Capabilities and Recent Advances. *International Journal of Network Management* 27 (5): e1973.

Hossain, Md F., Ayman U. Mahin, T. Debnath, Farjana B. Mosharrof, and Khondoker Z. Islam. 2019. Recent Research in Cloud Radio Access Network (C-RAN) for 5G Cellular Systems-A Survey. *Journal of Network and Computer Applications* 139: 31–48.

Jahan, R. 2015. Unlocking the True Potential of 5G: Techniques for Latency Reduction.

Li, Shancang, Li Da Xu, and Shanshan Zhao. 2018. 5G Internet of Things: A Survey. *Journal of Industrial Information Integration* 10: 1–9.

Liu, C. 2017. *Architectural Evolution and Novel Design of Fiber-Wireless Access Networks.* Springer.

Lynn, T., A. Gourinovitch, S. Svorobej, and P. Endo. 2018. Software Defined Networking and Network Functions Virtualization. Accessed 19 July 2019. https://recap-project.eu/news/software-defined-networking-and-network-functions-virtualization/

Marotta, Marcelo A., M. Kist, Juliano A. Wickboldt, Lisandro Z. Granville, J. Rochol, and Cristiano B. Both. 2017. Design Considerations for Software-Defined Wireless Networking in Heterogeneous Cloud Radio Access Networks. *Journal of Internet Services and Applications* 8 (1): 18.

Miozzo, M., N. Bartzoudis, M. Requena, O. Font-Bach, P. Harbanau, D. López-Bueno, M. Payaró, and J. Mangues. 2018. *SDR and NFV Extensions in the NS-3 LTE Module for 5G Rapid Prototyping.* IEEE Wireless Communications and Networking Conference (WCNC), June.

Mishra, A. 2018. *Fundamentals of Network Planning and Optimisation 2G/3G/4G: Evolution to 5G.* India: Wiley. Accessed 10 March 2019. https://books.google.

com.br/books?id=pbpmDwAAQBAJ&printsec=frontcover&hl=pt-BR&source=gbs_ge_summary_r&cad=0#v=onepage&q&f=false

Network Functions Virtualisation (NFV). 2013. Use Cases, ETSI, Sophia Antipolis, France.

Ordonez-Lucena, J., P. Ameigeiras, D. Lopez, Juan J. Ramos-Munoz, J. Lorca, and J. Folgueira. 2017. Network Slicing for 5G with SDN/NFV: Concepts, Architectures, and Challenges. *IEEE Communications Magazine* 55 (5): 80–87.

Pan, C., M. Elkashlan, J. Wang, J. Yuan, and L. Hanzo. 2018. User-Centric C-RAN Architecture for Ultra-dense 5G Networks: Challenges and Methodologies. *IEEE Communications Magazine* 56: 14–20.

Peña, M.A.L., Hector Humanes, J. Forsman, T. Le Duc, P. Willis, and M. Noya. 2019. Case Studies in Application Placement and Infrastructure Optimisation.

Peng, M., Y. Liu, D. Wei, W. Wang, and H.H. Chen. 2011. Hierarchical Cooperative Relay based Heterogeneous Networks. *IEEE Wireless Communications* 18 (3): 48–56.

Peng, M., Y. Li, Z. Zhao, and C. Wang. 2015a. System Architecture and Key Technologies for 5G Heterogeneous Cloud Radio Access Networks. *IEEE Network* 29 (2): 6–14.

Peng, M., C. Wang, V. Lau, and H.V. Poor. 2015b. Fronthaul-Constrained Cloud Radio Access Networks: Insights and Challenges. *arXiv preprint arXiv:1503.01187.*

Peng, M., Y. Sun, X. Li, Z. Mao, and C. Wang. 2016a. Recent Advances in Cloud Radio Access Networks: System Architectures, Key Techniques, and Open Issues. *IEEE Communications Surveys & Tutorials* 18 (3): 2282–2308.

Peng, M., S. Yan, K. Zhang, and C. Wang. 2016b. Fog Computing Based Radio Access Networks: Issues and Challenges. *IEEE Network* 30 (4): 46–53.

Popovski, Petar, Kasper Fløe Trillingsgaard, Osvaldo Simeone, and Giuseppe Durisi. 2018. 5G Wireless Network Slicing for eMBB, URLLC, and mMTC: A Communication-Theoretic View. *IEEE Access* 6: 55765–55779.

Ren, H., N. Liu, C. Pan, M. Elkashlan, A. Nallanathan, X. You, and L. Hanzo. 2018. Low-Latency C-RAN: An Next-Generation Wireless Approach. *IEEE Vehicular Technology Magazine*, April.

Rost, P., Carlos J. Bernardos, A. De Domenico, M. Di Girolamo, M. Lalam, A. Maeder, D. Sabella, and D. Wübben. 2014. Cloud Technologies for Flexible 5G Radio Access Networks. *IEEE Communications Magazine* 52 (5): 68–76.

Schiller, E., N. Nikaein, R. Favraud, and K. Kostas. 2015. *Network Store: Exploring Slicing in Future 5G Networks.* Paper presented at 10th ACM Workshop on Mobility in the Evolving Internet Architecture (ACM MobiArch). https://doi.org/10.1145/2795381.2795390.

Sun, Y., and M. Peng. 2018. Recent Advances of Heterogenous Radio Access Networks: A Survey. *Journal of Mobile Multimedia* 14 (4): 345–366.

Suryaprakash, V., R. Peter, and F. Gerhard. 2015. Are Heterogeneous Cloud-based Radio Access Networks Cost Effective? *IEEE Journal on Selected Areas in Communications* 33 (10): 2239–2251.

Taleb, T., and A. Ksentini. 2013. Follow Me Cloud: Interworking Federated Clouds and Distributed Mobile Networks. *IEEE Network* 27 (5): 12–19.

Taleb, T., A. Ksentini, and R. Jantti. 2016. "Anything as a Service" for 5G Mobile Systems. *IEEE Network* 30 (6): 84–91.

Tian, F., Z. Peng, and Y. Zheng. 2017. A Survey on C-RAN Security. *IEEE Access* 5: 13372–13386.

Tran, Huu Q., Phuc Q. Truong, Ca V. Phan, and Quoc-Tuan Vien. 2017. *On the Energy Efficiency of NOMA for Wireless Backhaul in Multi-tier Heterogeneous CRAN*. Paper presented at International Conference on Recent Advances in Signal Processing, Telecommunications & Computing (SigTelCom). IEEE. https://ieeexplore.ieee.org/document/7849827.

Wang, H., C. Wang, Derrick Wing K. Ng, H. Lee Moon, and J. Xiao. 2016. Artificial Noise Assisted Secure Transmission for Distributed Antenna Systems. *IEEE Transactions on Signal Processing* 64 (15): 4050–4064.

Wang, X., C. Cavdar, L. Wang, M. Tornatore, H.S. Chung, H.H. Lee, S.M. Park, and B. Mukherjee. 2017. Virtualized Cloud Radio Access Network for 5G Transport. *IEEE Communications Magazine* 55: 202–209.

Wang, K., Y. Kun, and Sarathchandra M. Chathura. 2018. Joint Energy Minimization and Resource Allocation in C-RAN with Mobile Cloud. *IEEE Transactions on Cloud Computing* 6: 760–770.

Wisely, D., Ning Wang, and Rahim Tafazolli. 2018. Capacity and Costs for 5G Networks in Dense Urban Areas. *IET Communications* 12 (19): 2502–2510.

Wu, J., Z. Zhang, Y. Hong, and Y. Wen. 2015. Cloud Radio Access Network (C-RAN): A Primer. *IEEE Network* 29: 35–41.

Yang, C., Z. Chen, B. Xia, and J. Wang. 2015. When ICN Meets C-RAN for HetNets: An SDN Approach. *IEEE Communications Magazine* 53 (11): 118–125.

Zeng, D., J. Zhang, S. Guo, L. Gu, and K. Wang. 2017. Take Renewable Energy into CRAN Toward Green Wireless Access Networks. *IEEE network* 31 (4): 62–68.

Zhang, Y., and Y. Wang. 2016. *A Framework for Energy Efficient Control in Heterogeneous Cloud Radio Access Networks*. Paper presented at IEEE/CIC International Conference on Communications in China (ICCC Workshops). IEEE. https://ieeexplore.ieee.org/document/7586713.

Zhang, J., W. Xie, and F. Yang. 2015. *An Architecture for 5G Mobile Network based on SDN and NFV*. Paper presented at 6th International Conference on Wireless, Mobile and Multi-Media (ICWMMN 2015). IEEE. https://ieeexplore.ieee.org/document/7453882.

Zhang, H., Y. Qiu, X. Chu, K. Long, and Victor C.M. Leung. 2017. Fog Radio Access Networks: Mobility Management, Interference Mitigation, and Resource Optimization. *IEEE Wireless Communications* 24 (6): 120–127.

Zhou, X., R. Li, T. Chen, and H. Zhang. 2016. Network Slicing as a Service: Enabling Enterprises' Own Software-Defined Cellular Networks. *IEEE Communications Magazine* 54 (7): 146–153.

CHAPTER 4

Orchestration from the Cloud to the Edge

Sergej Svorobej, Malika Bendechache, Frank Griesinger,
and Jörg Domaschka

Abstract The effective management of complex and heterogeneous computing environments is one of the biggest challenges that service and infrastructure providers are facing in the cloud-to-thing continuum era. Advanced orchestration systems are required to support the resource management of large-scale cloud data centres integrated with the big data generation of IoT devices. The orchestration system should be aware about all available resources and their current status in order to perform dynamic allocations and enable short time deployment of applications. This chapter will review the state of the art with regards to orchestration along the cloud-to-thing continuum with a specific emphasis on container-based orchestration (e.g. Docker Swarm and Kubernetes) and fog-specific orchestration architectures (e.g. SORTS, SOAFI, ETSI IGS MEC, and CONCERT).

S. Svorobej (✉) • M. Bendechache
Irish Institute of Digital Business, Dublin City University, Dublin, Ireland
e-mail: sergej.svorobej@dcu.ie; malika.bendechache@dcu.ie

F. Griesinger • J. Domaschka
Institute for Organisation and Management of Information Systems (OMI),
University of Ulm, Ulm, Germany
e-mail: frank.griesinger@uni-ulm.de; joerg.domaschka@uni-ulm.de

© The Author(s) 2020
T. Lynn et al. (eds.), *The Cloud-to-Thing Continuum*, Palgrave
Studies in Digital Business & Enabling Technologies,
https://doi.org/10.1007/978-3-030-41110-7_4

Keywords Cloud computing • Edge computing • Fog computing • Orchestration • Management • Container orchestration • Orchestration tools • Orchestration standards • Orchestration challenges • Orchestration architectures

4.1   INTRODUCTION

The inarguable success of cloud computing combined with rapid growth in adoption of Internet services is resulting in an unprecedented demand for computing resources. However, cloud computing performance for many applications depends closely on the network latency. In particular, the strength of network connectivity is crucial for large data sets. As more and more data is generated by enterprises and consumers, particularly with the adoption of the Internet of Things (IoT), traditional cloud connectivity may not be sufficient (Carnevale et al. 2018). To make up for the lack of speed and connectivity with conventional cloud computing, processing for mission-critical applications will need to occur closer to the data source. Processing the data close to where it originated is referred to as *edge computing* and *fog computing*.

Edge computing is pushing computing applications, data, and services away from centralised cloud data centre architectures to the edges of the underlying network (Barika et al. 2019). It is defined by NIST (Iorga et al. 2018) *"as a local computing at the network layer encompassing the smart end-devices and their users. It runs specific applications in a fixed logic location and provides a direct transmission service."* It promises to reduce the amount of data pushed to centralised cloud data centres avoiding load on the network and therefore is beneficial for analytics and knowledge-based services. Edge computing also leads to lower latencies, hence increasing communication velocity, reducing wider network footprints and avoiding congestion. As it reduces the distance the data must travel, it boosts the performance and reliability of latency-critical applications and services.

Service orchestration is an arrangement of auxiliary system components that cloud providers can use for coordination and management of computing resources to ensure service provision to cloud consumers (Bohn et al. 2011). Orchestration can also be defined as the use of programming technology to manage the interconnections and interactions among workloads on distributed edge-cloud infrastructure (Mahmoudi et al. 2018). This is accomplished through three main attributes of orchestration, which are closely related: service orchestration, workload orchestration, and resource orchestration. An orchestration platform usually integrates permission, checks for security, and compliance (Ranjan et al. 2015). Orchestration may also integrate components from various domains, for example provide connections between network-deployed components and fixed applications and resources. For some applications, the integration of virtualised components with the data centre is the only needed orchestration type.

Cloud-to-Edge orchestration is a crucial feature for many IT organisations and DevOps adopters as a way to speed the delivery of services, simplify optimisation, and reduce costs (Nygren et al. 2010). A cloud orchestrator automates the management, coordination, and organisation of distributed computer systems, services, and middleware. In addition to reduced personnel involvement, orchestration eliminates the potential for errors introduced into provisioning, scaling, or other cloud processes. Orchestration supports the delivery of cloud resources to customers and end users, including in a self-service model where users request resources without IT's involvement (Carnevale et al. 2018).

Major cloud providers, such as Microsoft and Google, as well as third-party vendors, provide tools for orchestration as part of their services (e.g. AWS Cloud Formation, Google Cloud Composer, Azure Automation). With orchestration, the overall goal is to ensure successful hosting and delivery of applications. Currently provided functionality is still lacking focus on Quality of Service (QoS) requirements, however meeting the QoS objectives of users will gain further importance in the future. Examples of QoS functional and non-functional attributes includes performance statistics, consistency, security, integrity, reliability, renting cost, scalability, availability, legal, and regulatory concerns (Pahl et al. 2019).

The rest of the chapter is organised as follows. The next section provides a summary overview of day to day challenges in Cloud-to-Edge orchestration. Next, we outline current industry standards for orchestration architectures and orchestration tools within their respective subsections. Finally, we conclude with some closing remarks on the topic.

4.2   ORCHESTRATION CHALLENGES

The orchestration of virtualised environments is challenging due to the scale, heterogeneity, and diversity of resource types and the uncertainties of the underlying cloud environment. The uncertainties arise from a number of factors including resource capacity demand (e.g. bandwidth and memory), failures (e.g. failure of a network link), user access pattern (e.g. number of users and location) and lifecycle activities of applications. In particular, cloud resource orchestration is challenging because applications are composed of multiple, heterogeneous software and hardware resources, which may have integration and interoperation dependencies (Barika et al. 2019).

Orchestration along the Cloud-to-Edge continuum adds another layer of complexity and challenges. In the cloud-to-thing era, applications as well as storage are geo-distributed. Therefore, applications will need to be restructured to distribute logic across the network. Storage will likewise need to be decentralised. This creates new issues of reliability and data integrity that are inherent in broadly decentralised networks. Cloud servers become control nodes for intelligent edge devices, performing summary analytics while leaving real-time decision making to edge servers (Jiang et al. 2018). Therefore, there is a need for comprehensive orchestration techniques that can coordinate and schedule network services simultaneously through different technologies across the Cloud-to-Edge network (Vaquero et al. 2019). Table 4.1 summarises the emerging orchestration needs in the edge/fog computing technologies and the corresponding requirements for each need.

In order to orchestrate distributed system as cloud-to-thing computing, new architecture needs to be defined taking into account the above edge orchestration needs and requirements.

The orchestration and management of a cloud-to-thing architecture is mostly realised through virtualisation. As discussed in Chap. 2, the evolution of virtualisation has moved away from virtual machines towards more lightweight solutions such as containers. This is specifically relevant for application packaging at a software platform and application level. Different application packages such as containers have been proposed to cluster Cloud-to-Edge and solutions such as Docker container and Kubernetes architectures. Yet, there is still a need for a topology specification and a derived orchestration plan for cloud edge computing.

Table 4.1 Emerging orchestration needs in edge/fog computing. (Adapted from Vaquero et al. (2019))

Functional orchestration needs	Requirements per need
Dynamic coalitions of edge devices and cloudlets	• Locality-awareness • Dynamism • Churn • Scalability • Replacement • Recovery
Going beyond shadow devices for reliability. Dynamic end-to-end service availability.	• Device churn • Multi-tenant • Multi-domain
Smaller execution units, smaller state	• Larger scale • Finer grain
Diversity	• Heterogeneity
M2M confidentiality, wireless-based attacks, trust management	• Security
AAA, privacy-leakage	• Privacy
Ensure quality-of-service on a variety of infrastructure elements	• Heterogeneity • Multi-domain

4.3   INDUSTRY STANDARDS

4.3.1   Network Function Virtualisation

Network Function Virtualisation (NFV) is a constantly evolving paradigm which enables the virtualisation of chains of communication services thus replacing purpose-built hardware appliances. With the increase in network traffic diversity and capacity growth in 5G NFV concept offers greater degree of flexibility for network, cloud, and mobile service providers (Barakabitze et al. 2019). The benefits of virtualisation include scalability, elasticity, and cost savings to the service; however the management of NFV chains becomes a challenge. The European Telecommunications Standards Institute (ETSI) Industry Specification Group for NFV (ETSI ISG NFV) has proposed an Open Source NFV Management and Orchestration (MANO) framework which provides NFV operators with the standard tools and framework for NFV orchestration (ETSI 2019). The NFV-MANO architecture is defined by three main functional blocks (ETSI 2014):

- VNF Manager (VNFM)
- VNF Orchestrator (VNFO)
- Virtualised Infrastructure Manager (VIM)

The VNFM is responsible for the lifecycle management of the VNF instances such as image template instantiation, software upgrades, scaling, and instance termination. The VNFO is responsible for orchestrating numerous VIMs to fulfil more complex function objectives across multiple VNF groups. Finally, VIM is an interface for a single infrastructure domain that is responsible for control and management of resources such as computation, storage, and network at that particular location. The latest implementation of Open Source Mano (OSM) Release 6 deploys the framework as a cohort of configurable Docker containers which provide VNF management capabilities and can integrate with multiple VIMs using plugins.

In a bid to bring unity to the NFV environment, the Open Platform for NFV (OPNFV) was launched through Linux Foundation (OPNFV 2019). The OPNFV project goal is to establish an ecosystem for NFV solutions that integrates through joint collaboration of development and testing. The OPNFV is a midstream project that drives new features based on the upstream user feedback, and also ensures component continuous integration downstream through composition deployment and testing (Brockners 2016).

4.3.2   OpenFog Reference Architecture

To standardise and promote the use of the fog computing paradigm across various disciplines the OpenFog consortium[1] was founded by the industry and academia in the telecommunication field. The OpenFog consortium working group created the OpenFog Reference Architecture (RA) for fog computing which was adopted by the IEEE Standards Association (OpenFog Consortium 2018). The reference architecture provides an overview of fog opportunity areas, use cases, and introduces eight pillars of OpenFog RA:

- Security—trust, attestation, privacy
- Scalability—localised command control and processing, orchestration and analytics, avoidance of network taxes

---

[1] Merged with Industrial Internet Consortium in January 2019

- Openness—resource visibility and control, white box decision making, interop and data normalisation
- Autonomy—flexible, cognition and agility, value of data
- Programmability—programmable SW/HW, virtualisation and multi-tenant, app fluidity
- RAS—Reliability, Availability, and Serviceability
- Agility—tactical and strategic decision making, data to wisdom
- Hierarchy—fully cloud enabled, computational and system, autonomy at all levels

The pillars provide guidance and describe requirements for hardware manufacturers, software developers, system vendors, and other parties in the fog supply chain. This view aligns well with the ISO/IEC CD 30141 that defines an Internet of Things RA (International Organization for Standardization 2018). It points out several necessary capabilities of IoT systems including the realisation of automated network management; ensuring of maintainability over long periods of time and large geographical region, including the need for configuration changes; reliability, and resilience of the system; and the need for availability and therefore scalability. The realisation of all of these capabilities requires a huge degree of automation and hence, are well-suited for the use of an orchestrator.

4.3.3   Orchestration Architectures

Multiple resource orchestration and provisioning architectures were developed to take advantage of Cloud-to-Edge infrastructure and its features. Munoz et al. (2015) present a management and orchestration architecture based on Software Defined Networking (SDN) and NFV. This architecture allows dynamic deployment of virtual tenant networks (VTN) and required corresponding SDN controllers in distributed data centre network as NFVs. The proposed solution is compatible with NFV MANO and consists of the following main functional blocks: *Multidomain SDN Orchestrator, Multidomain Network Hypervisor, Intra-DC Cloud and Network Orchestrator* and *Global Cloud and Network Orchestrator.*

The *Multidomain SDN Orchestrator* mechanism acts as a "controller of the controllers" of end-to-end provisioning services using Control Orchestration Protocol (COP). It orchestrates services across heterogeneous network layer components at a higher abstraction level thus supporting multiple lower level technologies. The *Multidomain Network*

Hypervisor aggregates and partitions physical network resources into virtual resources forming multiple connections among VTNs. The network hypervisor can dynamically create, change, and delete network resources based on matrix of QoS requirements. The *Intra-DC Cloud and Network Orchestrator* is responsible for VM lifecycle management, that is creation, migration, and deletion within a data centre. In a distributed data centre network, there is a need for an integrated orchestration. The *Global Cloud and Network Orchestrator* architecture component is responsible for global network and resource provisioning. It ensures VM migration and end-to-end connectivity links setup between distributed data centre site locations. The Integrated SDN/NFV Management and Orchestration Architecture was validated by an implementation that was deployed across three data centres in Spain and Poland.

Yannuzzi et al. (2017) propose a novel converged architecture called the Digital IoT Fabric, that complies with both OpenFog and MANO standards. The design of the Digital IoT Fabric aims to deliver a uniform management to NFV and IoT services with the deployment options from cloud to edge. The architecture is logically separated into four components:

1. the sensors, actuators, and control layer;
2. the system view of hardware resources and software view of virtualisation layer components;
3. five perspectives that comprise of platform capabilities i.e. manageability, security, performance and scale, data analytics and control, IT business and cross-fog applications; and
4. the User Interface (UI) and cloud and OpenFog services layer.

The logical connection between OpenFog and VNF MANO is achieved through the link between the OpenFog Node Management component and the MANO VIM component, both of which manage virtual functions and virtual infrastructures. Yannuzzi et al. (2017) argue that such architecture allows automated orchestration across the Edge-to-Cloud continuum and can play a key role in merging of operational technology and information technology.

SmartFog is another novel fog architecture which was designed to resemble human brain functions, where fog devices and network communication channels are analogous to neurons and synapses (Kimovski et al. 2018). This nature-inspired fog architecture makes use of graph theory, machine learning, and multi-criteria decision making to make fast decisions and architecture structuring. The architecture enables self-clustering of fog devices based on functional areas, further extending parallels with nature, for example temperature sensors forming a group of thermoreceptors or camera sensors forming a group of photoreceptors. The proposed architectural model can be logically divided into three distinctive layers: cloud layer, fog layer, and IoT layer. The cloud layer is the top layer where IoT application components are deployed and governed by functional requirements. The fog layer is the intermediary tier between the cloud and IoT layers, where the SmartFog architecture evolves around. SmartFog manages fog layer resources available within fog devices to create data transmission and processing paths through establishing communication gateways and assigning resources needed to host IoT application components and temporary storage blocks. The spectral clustering approach is applied to the lower IoT layer to classify and group fog devices based on their functional resemblance. Such groups are then connected to cloud applications in the upper cloud layer through dynamic communication gateways in the intermediary tier of the fog layer. The SmartFog architecture concept was validated via simulation only and as such, remains only a theoretical contribution.

Velasquez et al. (2017) recognise the difference between cloud computing and fog computing requirements and propose the Supporting the Orchestration of Resilient and Trustworthy Fog Services (SORTS) framework which introduces new mechanisms for services and resources orchestration specifically in fog environment. SORTS aims to maintain acceptable levels of QoS through ensuring resilience, trustworthiness and low latency within the dynamicity of a fog environment. The framework proposes a hybrid approach by using service orchestration and choreography management approaches. The orchestration is defined as a centralised management mechanism for cloud level resource management in the upper tier of the architecture. While the choreography mechanism is dedicated to the lower architecture tier covering management of IoT device virtual clusters and fog instances. Such operational decoupling in management levels allows quicker reaction to the changes to virtual clusters without intervention of higher level service management.

A Service Orchestration Architecture for Fog-enabled Infrastructures (SOAFI) is proposed by de Brito et al. (2017) which is based on the core requirements of fog computing focusing on heterogeneity and dynamics of IoT devices. Authors of SOAFI consider every exposed computer interface as a resource, and therefore in control by a resource manager: resource examples include microservices, sensors CPU, memory, network, VMs, accelerators. The framework itself is split in two tiers *Fog Orchestrator* (FO) and *Fog Agent (FA)*. The FO manages the infrastructure of connected fog nodes; it keeps a database of available resources through a built discovery service. The FA is running on a fog node and provides monitoring and local access to resource management through the interface to the FO. The authors were successful in implementing and initial working prototype of SOAFI which was deployed in their IoT testbed.

A Cloud-Based Architecture for Next-Generation Cellular Systems, named CONCERT is proposed by Jingchu Liu et al. (2015). As the name suggests the architecture is targeted for management of cellular edge infrastructure embracing NFV services. The CONCERT approach is based on the concept of control and data plane decoupling where data plane embodies physical resources of edge infrastructure and the control plane coordinates physical resources through virtualisation. In addition, CONCERT allows physical resource placement and task scheduling in a bit for better service orchestration. The control plane entity, called *Conductor*, is at the centre of the proposed architecture design. It orchestrates and virtualises data plane resources as well as controlling software defined switches through centralised packet forwarding tables. This way the *Conductor* manages physical data plane resources as a central entity by provisioning them to a required VNF.

In an effort to bridge the gap between theory and practice, Santos et al. (2017) propose a container-based fog computing orchestration architecture. The proposed architecture was implemented using Kubernetes, an open source management solution for containerised applications, which was extended with network-aware scheduling (NAS) (Santos et al. 2019) and Integer Linear Programming (ILP) decision support for IoT service placement (Santos et al. 2017). The network-aware scheduling makes resource provisioning decisions by taking in consideration current load status of available network infrastructure and the target location of service. The ILP ensures close placement proximity of IoT application services to the end devices which use these services. The smart city scenario-based experiments show a 70% network latency reduction compared to the default Kubernetes scheduling setup with 1.22 ms scheduling decision time overhead.

4.3.4   Orchestration Tools

Industry standard and proposed orchestration architectures define high level system design best practices for multiple integrated functional components. However, in order to use any of the system design features in real world scenarios, an actual implementation has to take place. A wide range of resource management tools are available to orchestrate Cloud-to-Edge infrastructure which are outlined below.

Since edge site resources are considered to be limited due to constraints in physical hosting space, we focus primarily on container-supporting tools as containers have leaner resource overhead profiles when compared to virtual machines. A container is technology that provides lightweight virtualisation at the kernel level. It is a packaged, self-contained, ready-to-deploy set of parts of applications, that might include middleware and business logic in the form of binaries and libraries to run the applications (Pahl and Lee 2015). Containers address concerns at the cloud PaaS level allowing to spawn self-contained applications on demand. Containers are often called building blocks of PaaS due to flexibility to be spawned on both physical and virtual infrastructures. Containers also relate to the IaaS level through sharing and isolation aspects that exemplify the evolution of OS and virtualisation technology. Docker is one of the most popular container tools for the Linux and Windows operating system with about 83% of market share followed by CoreOS rkt (12%), Mesos Containeriser (4%) and Li Linux Containers (LXC) (1%) (Carter 2018). Dockers are frameworks built around container engines (Turnbull 2014). They make containers a portable way to package applications to run in containers. The Open Container Initiative[2] is making a push to create de-facto standards for container runtime and image formats. In terms of a tiered application, a tier can be represented by a single container or a number of containers depending on application design and requirements.

Kubernetes is a popular open-source platform for managing containers and their hosted services. Kubernetes was initially developed by Google and open sourced in 2014; it is maintained by the Cloud Native Computing Foundation. Kubernetes is a modular platform that focuses on automation of container management tasks such as service discovery and load balancing, storage orchestration, deployment roll outs and rollbacks, container bin packing, self-healing, secret and configuration management. The

---

[2] https://www.opencontainers.org/

architecture of Kubernetes is divided into three distinct component areas—Master Components, Node Components, and Add-ons. The Master Components form the control plane of the cluster making global decisions on scheduling, backup and ensuring node pod deployments to hardware nodes. The Node Components form and maintain usable Kubernetes environment on a hardware node. These components are deployed on each individual node in the data centre that are zoned to be used for container hosting providing network proxy features, healthy container state and enable container runtime features. Add-ons are an optional component group that complements the Master and Node Components by providing additional DNS, web UI and resource monitoring features (Kubernetes 2019). Kubernetes is used as a base platform for Red Hat OpenShift[3] and Rancher,[4] which provide additional features for Kubernetes cluster management, resource provisioning, monitoring and security. Recently, Rancher released K3s specifically tailored towards low-end infrastructure such as IoT gateways and extreme edge devices.

The Nebula Container Orchestrator[5] is an open-source project designed to manage large-scale clusters of Docker containers. The solution is specifically targeted at large-scale scenarios such as IoT devices or virtual Content Delivery Networks (vCDN) running Docker containers. The Nebula Container Orchestrator provides a REST API that can be used for sending management instructions to the deployed container groups such as rolling out updates, mounting volumes, changing images, monitoring health and performance and adjusting resource allocation. The architecture consists of two core components, *Manager* and *Worker*, and an optional monitoring component, *Reporter*. First the IoT device connects to the *manager* to retrieve group configuration information, and after configuration obtained the *worker* is handling device further. All of the architecture components are using a single scalable backend database (i.e. MongoDB[6]) to store configuration states and monitoring data. *Manager* is a fully stateless component that serves as an API endpoint to control the system. *Worker* is running on the remote container and manages the *Worker* by periodically pulling instructions from manager components. The *Reporter* component is used for collecting the data from the

---

[3] https://www.openshift.com/
[4] https://rancher.com/
[5] https://nebula-orchestrator.github.io/
[6] https://www.mongodb.com/

individual containers in the group to provide monitoring data to the system administrator. The Nebula Container Orchestrator is designed with scale in mind, and each component can be scaled to meet the demands of the system. Stress test results suggest a linear increase in number of IoT devices a single *Manager* component can handle from 7780 devices checking manager every 10 seconds to 466,800 devices checking every 600 seconds (Nebula 2019). Since multiple *Manager* components can be dynamically deployed the orchestrator provides a flexible solution for large-scale containerised service deployments.

Swarm is an open-source native container orchestration engine build for the Docker platform. The Docker integration allows Docker Engine CLI to be used directly to issue Swarm commands providing streamlined Docker container cluster management experience. Since the Swarm mode is already a part of the Docker engine, no other additional orchestration tools are needed when working with Docker-based containers. The Swarm architecture consists of manager nodes, distributed state store, and worker nodes (Docker Inc. 2019a). The manager nodes are responsible for maintaining cluster state, schedule services and provide access to the functionality over web API endpoints. It is recommended to run multiple managers as a safeguard against failures in order to maintain consistency of the entire swarm. The manager nodes use Raft (Ongaro and Ousterhout 2014) consensus algorithm for managing replicated logs via the internal distributed state store where each manager is connected. The worker nodes' sole purpose is to execute containers. Worker nodes do not use distributed state storage and don't provide services of manager nodes; however, a worker can be promoted to a manager with a single "*promote*" command as they are also instances of Docker Engine. This functionality is useful for node maintenance and failure recovery scenarios. Swarm includes features as incremental node updates, TLS-based authentication and traffic encryption, internal load balancing specification, and an API to connect external load balancers with support for the overlay networking and scaling (Docker Inc. 2019b).

4.4 Conclusion

The last decade has brought a rapid emergence of smart devices which encouraged the development of cloud computing, hardware, networks, and mobility. Both the enterprise and consumer landscape are seeing an increase in these device numbers as the value of rapid information access is being realised in day-to-day scenarios. The devices are used in geographically remote locations where access to Internet connection as well as the remote cloud services, is not stable, hence the need to process generated data locally. This introduces an additional unique layer of heterogeneity with physical form factor variability as well as unique network data transfer capability. Meanwhile the increasing consumer and enterprise service demand is creating significant strain on the telecommunication compute and network infrastructure. Hardware heterogeneity, scalability and latency are some of the main challenges that Cloud-to-Edge infrastructure providers are facing on a day-to-day basis in order to uphold QoS that are expected by customers.

Orchestrators on the other hand have emerged together with cloud computing and provide a mature approach to coordinate the automated managing tasks for distributed applications running on IaaS or container-based environments. The resource orchestration approach and tools stack play an important role in distributed service delivery. Considering that edge and fog applications need to deal with more dynamic and less predictable environments, their operators are even more dependent on reliable and efficient orchestrators that need to handle the new challenges: the use of geo-distributed infrastructure demands for more detailed understanding of application behaviour; support for federation, as there is a high chance that edge environments will span multiple providers.

There is currently a strong movement to establish cloud and fog computing as business models and a movement towards fog orchestrators. Also, multiple active standardisation initiatives exist. Nevertheless, this chapter showed that the current state of the art in Cloud-to-Edge orchestration does not address all challenges and that more work in research and standardisation needs to be done. Just as the paradigm of cloud-native applications has given momentum to the development of cloud orchestrators, establishing a commonly accepted definition of fog-native applications might accelerate the evolvement of fog orchestrators.

References

Barakabitze, Alcardo Alex, Lingfen Sun, Is-Haka Mkwawa, and Emmanuel Ifeachor. 2019. A Novel QoE-Aware SDN-Enabled, NFV-based Management Architecture for Future Multimedia Applications on 5G Systems. *arXiv preprint arXiv:1904.09917*, April. http://arxiv.org/abs/1904.09917.

Barika, Mutaz, Saurabh Garg, Albert Y. Zomaya, Lizhe Wang, A. van Moorsel, and Rajiv Ranjan. 2019. Orchestrating Big Data Analysis Workflows in the Cloud: Research Challenges, Survey, and Future Directions. *ACM Computing Surveys* 52: 1–37.

Bohn, Robert B., John Messina, Fang Liu, Jin Tong, and Jian Mao. 2011. *NIST Cloud Computing Reference Architecture*. 2011 IEEE World Congress on Services, 594–596. IEEE. https://doi.org/10.1109/SERVICES.2011.105.

Brito, Mathias Santos de, Saiful Hoque, Thomas Magedanz, Ronald Steinke, Alexander Willner, Daniel Nehls, Oliver Keils, and Florian Schreiner. 2017. *A Service Orchestration Architecture for Fog-Enabled Infrastructures*. 2017 Second International Conference on Fog and Mobile Edge Computing (FMEC), 127–32. IEEE. https://doi.org/10.1109/FMEC.2017.7946419.

Brockners, Frank. 2016. What is OPNFV? OPNFV Summit.

Carnevale, Lorenzo, Antonio Celesti, Antonino Galletta, Schahram Dustdar, and Massimo Villari. 2018. *From the Cloud to Edge and IoT: A Smart Orchestration Architecture for Enabling Osmotic Computing*. 2018 32nd International Conference on Advanced Information Networking and Applications Workshops (WAINA), 2018-January, 419–424. IEEE. https://doi.org/10.1109/WAINA.2018.00122.

Carter, Eric. 2018. Docker Usage Report. https://sysdig.com/blog/2018-docker-usage-report/.

Docker Inc. 2019a. How Nodes Work | Docker Documentation. https://docs.docker.com/engine/swarm/how-swarm-mode-works/nodes/.

———. 2019b. Swarm Mode Overview | Docker Documentation. https://docs.docker.com/engine/swarm/.

ETSI. 2014. Network Functions Virtualisation (NFV); Management and Orchestration. *Etsi*, vol. 2. http://www.embase.com/search/results?subaction=viewrecord&from=export&id=L34467660.

———. 2019. ETSI—Open Source Mano | Open Source Solutions | Mano NFV. https://www.etsi.org/technologies/nfv/open-source-mano.

International Organization for Standardization. 2018. *Information Technology—Internet of Things Reference Architecture (IoT RA)*. ISO/IEC130141:20182018, Geneva. https://www.iso.org/standard/65695.html.

Iorga, Michaela, Larry Feldman, Robert Barton, Michael J. Martin, Nedim S. Goren, and Charif Mahmoudi. 2018. Fog Computing Conceptual Model.

Jiang, Yuxuan, Zhe Huang, and Danny H.K. Tsang. 2018. Challenges and Solutions in Fog Computing Orchestration. *IEEE Network* 32 (3): 122–129. https://doi.org/10.1109/MNET.2017.1700271.

Kimovski, Dragi, Humaira Ijaz, Nishant Saurabh, and Radu Prodan. 2018. *Adaptive Nature-Inspired Fog Architecture*. 2018 IEEE 2nd International Conference on Fog and Edge Computing (ICFEC), 1–8. IEEE. https://doi.org/10.1109/CFEC.2018.8358723.

Kubernetes. 2019. Kubernetes. https://kubernetes.io.

Liu, Jingchu, Tao Zhao, Sheng Zhou, Cheng Yu, and Zhisheng Niu. 2015. CONCERT: A Cloud-based Architecture for Next-Generation Cellular Systems. *IEEE Wireless Communications* 21 (6): 14–22. https://doi.org/10.1109/mwc.2014.7000967.

Mahmoudi, Charif, Fabrice Mourlin, and Abdella Battou. 2018. *Formal Definition of Edge Computing: An Emphasis on Mobile Cloud and IoT Composition.* 2018 Third International Conference on Fog and Mobile Edge Computing (FMEC), 34–42. IEEE. https://doi.org/10.1109/FMEC.2018.8364042.

Muñoz, Raul, Ricard Vilalta, Ramon Casellas, Ricardo Martinez, Thomas Szyrkowiec, Achim Autenrieth, Víctor López, and Diego López. "Integrated SDN/NFV management and orchestration architecture for dynamic deployment of virtual SDN control instances for virtual tenant networks." Journal of Optical Communications and Networking 7, no. 11 (2015): B62-B70.

Nebula. 2019. Scaling—Nebula Container Orchestrator. https://nebula.readthedocs.io/en/latest/scaling/.

Nygren, Erik, Ramesh K. Sitaraman, and Jennifer Sun. 2010. The Akamai Network. *ACM SIGOPS Operating Systems Review* 44 (3): 2. https://doi.org/10.1145/1842733.1842736.

Ongaro, Diego, and John Ousterhout. 2014. *In Search of an Understandable Consensus Algorithm.* 2014 Annual Technical Conference, 305–319.

OpenFog Consortium. 2018. *IEEE Standard for Adoption of OpenFog Reference Architecture for Fog Computing.* IEEE Std 1934–2018, August, 1–176. https://doi.org/10.1109/IEEESTD.2018.8423800.

OPNFV. 2019. Software—OPNFV. https://www.opnfv.org/software.

Pahl, Claus, and Brian Lee. 2015. *Containers and Clusters for Edge Cloud Architectures—A Technology Review.* 2015 3rd International Conference on Future Internet of Things and Cloud, 379–386. IEEE. https://doi.org/10.1109/FiCloud.2015.35.

Pahl, Claus, Antonio Brogi, Jacopo Soldani, and Pooyan Jamshidi. 2019. Cloud Container Technologies: A State-of-the-Art Review. *IEEE Transactions on Cloud Computing* 7 (3): 677–692. https://doi.org/10.1109/TCC.2017.2702586.

Ranjan, Rajiv, Boualem Benatallah, Schahram Dustdar, and Michael P. Papazoglou. 2015. Cloud Resource Orchestration Programming: Overview, Issues, and Directions. *IEEE Internet Computing* 19 (5): 46–56. https://doi.org/10.1109/MIC.2015.20.

Santos, Jose, Tim Wauters, Bruno Volckaert, and Filip De Turck. 2017. *Resource Provisioning for IoT Application Services in Smart Cities.* 2017 13th International Conference on Network and Service Management (CNSM), 2018-January, 1–9. IEEE. https://doi.org/10.23919/CNSM.2017.8255974.

———. 2019. *Towards Network-Aware Resource Provisioning in Kubernetes for Fog Computing Applications.* IEEE Conference on Network Softwarization (NETSOFT), Paris.

Turnbull, James. 2014. *The Docker Book: Containerization Is the New Virtualization.* James Turnbull.

Vaquero, Luis M., Felix Cuadrado, Yehia Elkhatib, Jorge Bernal-Bernabe, Satish N. Srirama, and Mohamed Faten Zhani. 2019. Research Challenges in Nextgen Service Orchestration. *Future Generation Computer Systems* 90: 20–38. https://doi.org/10.1016/j.future.2018.07.039.

Velasquez, Karima, David Perez Abreu, Diogo Goncalves, Luiz Bittencourt, Marilia Curado, Edmundo Monteiro, and Edmundo Madeira. 2017. *Service Orchestration in Fog Environments*. Proceedings—2017 IEEE 5th International Conference on Future Internet of Things and Cloud, FiCloud 2017, 2017-January, 329–336. https://doi.org/10.1109/FiCloud.2017.49.

Yannuzzi, M., R. Irons-Mclean, F. van Lingen, S. Raghav, A. Somaraju, C. Byers, T. Zhang, et al. 2017. *Toward a Converged OpenFog and ETSI MANO Architecture*. 2017 IEEE Fog World Congress (FWC), 1–6. IEEE. https://doi.org/10.1109/FWC.2017.8368535.

CHAPTER 5

Living at the Edge? Optimizing availability in IoT

Guto L. Santos, Kayo H. de C. Monteiro,
and Patricia Takako Endo

Abstract Cloud, edge, and fog computing enable Internet of Things (IoT) applications, offering high connectivity, scalability, and high availability. Smart cities, smart agriculture, and e-health systems are examples of IoT applications that can exploit the opportunities generated by these technologies. However, due to the scale and complexity of the IoT and the heterogeneity of the devices and service level expectations, resource management is not a trivial task. While facilitating storage and processing

G. L. Santos (✉)
Universidade Federal de Pernambuco, Recife, Brazil
e-mail: guto.leoni@gprt.ufpe.br

K. H. de C. Monteiro
Universidade de Pernambuco, Recife, Brazil
e-mail: khcm@ecomp.poli.br

P. T. Endo
Universidade de Pernambuco, Recife, Brazil

Irish Institute of Digital Business, Dublin City University, Dublin, Ireland
e-mail: patricia.endo@upe.br

© The Author(s) 2020
T. Lynn et al. (eds.), *The Cloud-to-Thing Continuum*, Palgrave
Studies in Digital Business & Enabling Technologies,
https://doi.org/10.1007/978-3-030-41110-7_5

at the end device (the edge), at the intermediary layer (the fog), or centrally (the cloud), new points of failure are introduced at and between each layer. In some use cases, such as e-health, device availability also has high criticality. Any downtime impacting one or more components in the architecture can result in adverse effects and/or additional logistical effort and cost. This chapter discusses extant research on how cloud, fog, and edge computing is being used in smart city, smart agriculture, and e-health systems.

Keywords Availability • Smart city • Smart agriculture • e-health • Internet of Things • Edge computing • Cloud computing • Fog computing

5.1   Introduction

For large-scale Internet of Things use cases, such as smart cities, cloud computing offers a flexible and on-demand infrastructure to execute real-time analytics applications and context-specific information systems. In addition, it also provides support for data retrieval and visualization, which can provide relevant information for city managers to create new public policies and consequently improve the city operations (Kakderi et al. 2019). While cloud computing can mitigate some of the limitations of IoT devices, the location of cloud data centres can introduce issues for delay-sensitive applications. Usually, cloud infrastructures are located far away from the edge devices resulting in a high delay for both transmitting large volumes of data and providing a response (Tang et al. 2015). To solve this problem, fog computing is a technology that acts between the edge devices and cloud computing. Fog computing extends the cloud capabilities providing cloud features closer to the edge devices using low cost devices but with computational capabilities needed to support local data analysis (He et al. 2017; Santos et al. 2018).

Despite the advantages that cloud-to-thing (C2T) integration technologies bring to IoT applications, its adoption also brings challenges. In such scenarios, the system relies on different operational layers to offer a variety of applications to the end-user, increasing the complexity of resource management (da Silva Lisboa et al. 2018). Moreover, while improving the availability of the service as a whole, C2T solutions present more points of possible failure, for example in the IoT sensors, fog devices, and data centre subsystems. In several scenarios, system availability is critical due to strict requirements, such as real-time data health monitoring, for example patient vital data, and real-time decision making, for example rerouting due a transit accident.

This chapter discusses extant research on how cloud, fog, and edge computing is being used in smart city, smart agriculture, and the Internet of Medical Things.

5.2   Smart Cities

By 2030, more than 60% of the world's population will live in an urban environment (Gaur et al. 2015). This urban growth places significant strains on the city infrastructure. New solutions are required to mitigate emerging problems from rapid urbanization and an aging population such as energy consumption, safety, public services, sustainable development amongst others (Arroub et al. 2016). The use of information technologies is a solution to create smart urban environments to both reduce the impact of population growth and improve overall quality of life (Chakrabarty and Engels 2016).

A smart city is an urban space composed of several complex systems, covering infrastructures, technologies, social and political structures, and the economy (Gaur et al. 2015). To control and monitor this environment, a vast array of heterogeneous devices, such as street cameras for security systems and sensors for utility and transportation systems are placed across the city, which generate a huge amount of data (Arasteh et al. 2016). Such data can be used to identify possible bottlenecks and also to provide insights for city managers thereby supporting better decision making. It unsurprising therefore that the market for smart city solutions is forecast to grow to US$237.6 billion by 2025, expanding at a CAGR of 18.9% from 2019 to 2025 (Grand View Research 2019).

Notwithstanding the opportunities inherent in smart cities, they also bring several challenges. A smart city is a mixture of multiple systems with distinct, sometimes conflicting objectives. These systems may make use of their own (devices) end-points or share municipal devices to collect data over time, many of which have no significant computational capability to store and process this data. These simple devices act as the city's sensing layer located at the edge of the network.

An example of smart city infrastructure for traffic monitoring that integrates edge, fog and cloud computing is illustrated in Fig. 5.1. The cars and roadside units generate data about the number of cars driving on the

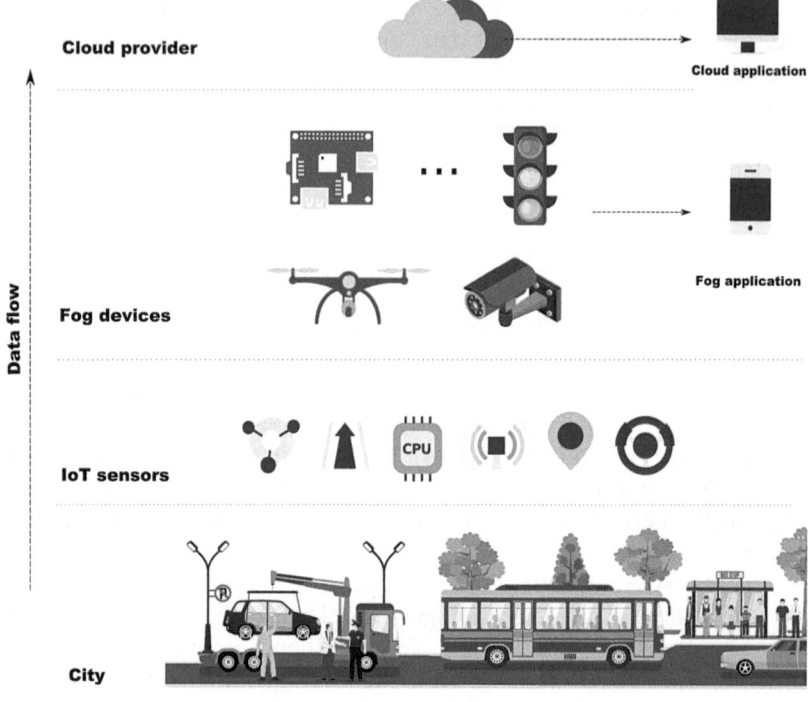

Fig. 5.1 Smart city use case with IoT sensors, fog devices, and cloud computing

streets, street conditions, and unexpected events, such as accidents. This information can be used to adjust the traffic lights in order to relive potential traffic congestions in and around the location of the incident. This information can be processed locally by the traffic lights (fog nodes) to reduce the delay. However, the cars and the city infrastructure could process incidents, for example car crashes, and send alerts to change the route for vehicles and give priority to city or emergency services to arrive at the scene of the incident (Ali and Ghazal 2017).

Integrating edge, fog, and cloud infrastructure to implement smart city services introduces high management complexity. Multiple devices and applications must work together to keep the services and data available for the applications. Generally, device failure or data loss can have a significant impact on critical systems and in smart cities, these problems can cause adverse effects in urban contexts (Baldoni et al. 2015). The infrastructure and massive volumes of data generated by smart cities, often requiring real-time analysis require robust and efficient resource planning and, consequently, better resource utilization (Bawany and Shamsi 2015).

The importance of the availability of data and services of smart cities have been addressed by some authors considering different smart city scenarios. For instance, smart parking solutions using IoT and cloud technologies are being proposed to help drivers and as a consequence decrease the carbon footprint. However, such solutions rely on specialized infrastructures (sensors) and real-time data availability.

Mugarza et al. (2019) address the need for high availability in smart city power management systems. They describe a solution for dynamic software updates for power system management. This type of system cannot be interrupted as any system outage or malfunction may result in a blackout in some or all of the city. The solution presented enables building power management software to be updated without restarting, thus ensuring that the system will be fully functioning.

The vehicular cloud, sometimes referred to as V2C, offers vehicular communication technologies and infrastructures for smart vehicles to communicate with other parties, among themselves and with the cloud (Soyturk et al. 2016). Smart-Vehicle-as-a-Service (SVaaS) acts as intermediary layer between the vehicular nodes (e.g. smart cars or municipal vehicles) and the service providers (e.g. municipal authorities or commercial services) so that relevant services such as transport management, on-demand transportation services, and media services, can be provided (Aloqaily et al. 2017). Smart vehicles is a good example of heterogeneity in end devices as smart cars may use a wide range of technologies, be used by multiple drivers, and have different levels of criticality depending on their use or importance as perceived by service or infrastructure providers. Aloqaily et al. (2017) present a system for discovering and selecting vehicular services based on QoE requirements and vehicle clusters that are geographically proximate and have similar service profiles.

As discussed, the provision of services and the availability of resources is a great challenge in smart city due to different demands, user requirements, and services specifications. For delay-sensitive applications, fog computing is a key technology since it provides cloud-like functionality close to the end-user thereby reducing the latency and increasing the application availability (Tang et al. 2015). In a smart city scenario, where a huge amount of data is generated, the fog nodes can act as a backup in case of a cloud failure, increasing the availability of the system as a whole (Stantchev et al. 2015). Finally, different redundancy mechanisms can be used between the fog applications and the cloud data centres. The edge nodes may provide more essential services in a distributed way, and in case of failures, other edge nodes can keep the service available.

5.3   SMART AGRICULTURE

Agriculture plays an important role in the world because as a source of livelihood and as its role in the global food supply chain. According to the WHO, an estimated 820 million people did not have enough to eat in 2018. The situation is most alarming in Africa, as the region has the highest rate of hunger in the world. In Eastern Africa, for instance, about 30.8% of the population is undernourished. Promoting the optimization of the food supply chain can help eradicate hunger and poverty in the world (Bu and Wang 2019) but also contribute significant to economic development (Gondchawar and Kawitkar 2016).

In recent years, digital monitoring and control are being used in agriculture systems to improve the food supply chain from farm to fork. Precision agriculture involves the measurement and optimization of granular field operations. Currently, sensors can offer highly accurate measurements of crop status, and based on those values, actuators are able to manage, for instance, irrigation, change climate factors, or enrich the soil with the needed nutrients (Mulla 2013). This, in turn, can result in improved harvest forecasting, collection, planning, and downstream operations (McKinsey 2016).

Again, smart agriculture bring challenges as well as opportunity. Smart agriculture is characterize by high volumes of heterogeneous sensors and actuators distributed across wide areas, often with intermittent connectivity. Figure 5.2 presents a smart agriculture use case with integrated IoT, fog, and cloud to increase process automation and efficiency. Heterogeneous sensors collect relevant crop data such as humidity, temperature, pH metering, and soil conditions; and heterogeneous actuators, such as water sprinklers, ventilation devices, lighting, automated windows (in glasshouses), and soil and water nutrition pumps react according to the data.

To overcome some of the coverage and connectivity issues inherent in agriculture, Zamora-Izquierdo et al. (2019) propose that at the crop layer (farm), IoT sensors and actuators are deployed and connected to fog devices. This connection can be made by the intermediary layer

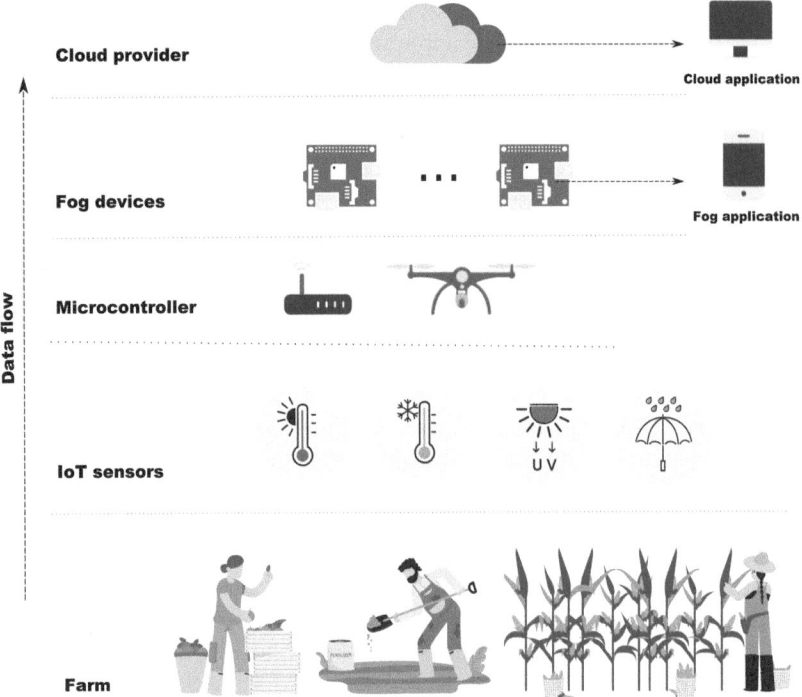

Fig. 5.2 Smart agriculture with integrated IoT sensors, microcontrollers, fog devices, and cloud computing

(represented by a microcontroller and/or a drone). These devices (e.g. Arduinos, ESP32, drones, etc.) have limited hardware capabilities to process and store the data from sensors, but have connectivity to forward data for superior layers. Fog devices are located geographically close to the IoT sensors and microcontrollers and can provide computational resources to process or storage data locally. Delay-sensitive applications can use fog devices to obtain fasters responses. Finally, cloud computing provides "unlimited" computational resources to process and store the data collected by sensors. Thus, further analysis that may demand more powerful resources, such as big data analysis and machine learning model training, may be executed on cloud environments. However, as cloud infrastructure is located far away from microcontrollers and fog devices, there may be a significant delay in data transmission. This may compromise the performance of the smart agriculture application.

Researchers have proposed a wide range of architectural solutions for smart agriculture. TongKe (2013) uses sensors to send information (such as water quality, monitoring of animal, and plant growth) directly to cloud data centres. This information is processed to provide relevant outputs for system users. Bawany and Shamsi (2015) propose a similar architecture where the information is sent for cloud and can be visualized using a smartphone app. As these architectures depend on the cloud for processing and storage data, a basic requirement is an Internet connection; this can be a constraint in remote places. Therefore, for smart agriculture scenarios, the usage of fog devices to process data locally can increase the availability of smart agriculture systems. In large scenarios, with a huge number of sensors and actuators, the combination of fog and cloud can be used to increase the coverage area of the system and improve the service and data availability.

A precision farming system using IoT and cloud infrastructure is presented by Ibrahim et al. (2018) using a horticulture use case, and specifically ornamental flowers. System inactivity can cause severe damage to crops, since ornamental flowers are very sensitive to humidity and temperature changes resulting in financial damage to the farmers. The proposed system is composed of a large number of sensors, controllers, and actuators, and a 1-in-2 fault-tolerant management system to mitigate the likelihood and impact of failures. By using the fault-tolerant approach, the system availability increased from 99.45% to 99.997%.

In smart agriculture, downtime can be very costly (Ibrahim et al. 2018). For instance, if a temperature change is perceived but action is delayed or no action is taken, it can cause significant financial losses because the entire crop can be compromised. Downtimes can occur due to different reasons, such as sensor and actuator failure. In these cases, minimizing the system downtime becomes critical. Ibrahim et al. (2018) suggest developing countries encounter high repair times as spare components are not usually stored on site and farmers may have to import them, and replacement costs are rarely predictable. Again linking, system failure and downstream replacement logistics could result in better mean time to repair metrics and mitigate financial losses.

5.4   The Internet of Medical Things

The promise of information technologies for health care has been a continuous theme in discourse by policymakers, academia, and industry. It is widely accepted that in the institution-centric health care system that dominates most developed economies, patients and health care professionals are separated from each other by time and space (Tachakra et al. 2003; Topol 2015). Time, place and people drive not only costs but inconvenience in to health care systems suffering under the increasing pressure of aging populations and chronic diseases (Topol 2015; Deloitte 2018).

Gatouillat et al. (2018) defined the Internet of Medical Things (IOMT) as the interconnection of medical-grade devices with broader health care infrastructures, connecting personal medical devices with each other and with health care providers whether hospitals, medical researchers, or private companies. The advantages of such a vision of e-health ubiquity and interconnectivity are multifold. As well as unburdening the health system, IoMT technologies can contribute to increased efficiency, reduced costs, risk mitigation, improved quality of care and quality of life, and increased transparency in the health system (Darkins et al. 2008; Ossebaard et al. 2013; Farahani et al. 2018). Despite these advantages, the IoMT is not without challenges, not least structural and technical ones. For example, the pervasiveness and criticality of e-health technologies, the impact of adverse outcomes, and sensitivity of personal health information requires IoMT systems to have higher levels of reliability, safety, and security than typical systems (Gatouillat et al. 2018).

For the most part, the IoMT is configured along the C2T continuum comprising three layers (Fig. 5.3): (1) cloud computing and big data infrastructure, (2) Internet-connected gateways, and (3) a (body) sensor or sensor networks (Rahmani et al. 2018). Advances in sensor design, cloud computing, and wireless communication technologies has made it possible to develop more affordable medical systems. In the last two decades, there has been a rapid advance in the sophistication of sensors including intelligent sensors, in vivo sensors, and sensors that can increasingly mimic the biological senses. The availability of low cost relatively sophisticated sensors for monitoring patient life plays a significant role in the evolution of the IoMT and vice-versa (Chiuchisan et al. 2014). A reliable system in the IoMT must achieve its functional goals at all times to avoid patient safety

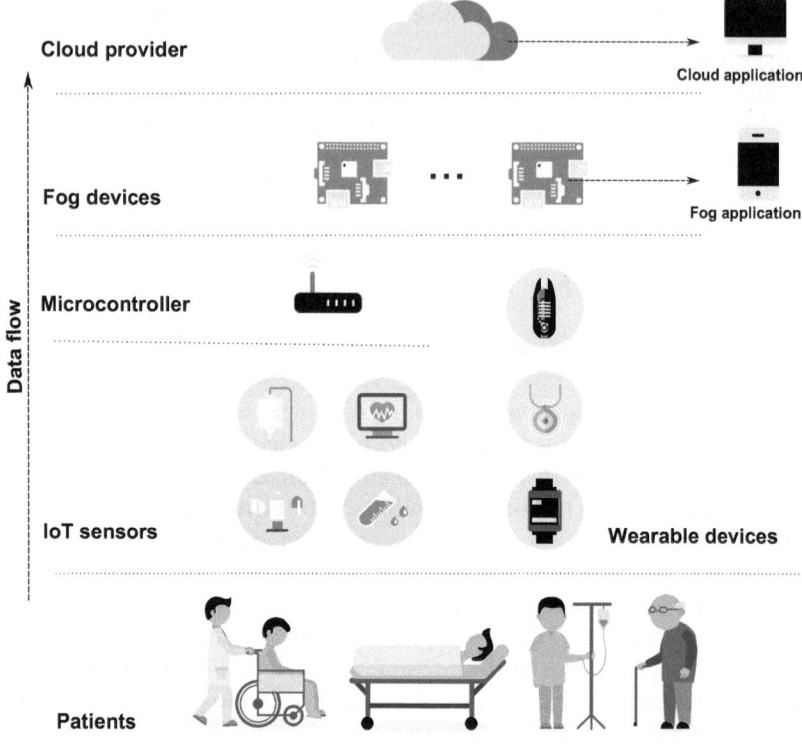

Fig. 5.3 Smart e-health scenario integrating IoT sensors and wearable devices, microcontrollers, fog devices, and cloud computing

issues, adverse outcomes, or additional effort and costs by patients or health care providers (Gatouillat et al. 2018). At the same time, budgetary constraints remain. The system architect therefore has a trade-off to resolve between system availability and cost.

Araujo et al. (2014) propose stochastic models, Stochastic Petri Nets (SPN), and Reliability Block Diagrams (RBD), to represent an e-health service relying on mobile cloud computing infrastructure, that is cloud infrastructure, wireless communications, and a mobile device. Availability analysis were conducted considering scenarios with different wireless communication channels (Wi-Fi and 4G), different battery discharge rates, and different timeout values. Results suggested that the timeout for delivering a message is the metric that had a greater impact on mHealth system availability, and that Wi-Fi and 4G communication channels presented similar impact.

Vargheese and Viniotis (2014) focus on usage of sensors to provide vital data as soon as the patient is discharged from the hospital, and they state that the availability of the cloud and IoT devices is a critical aspect in this scenario. However, not only do the devices (at software and hardware levels) play an important role regarding availability aspect; the role of the network connection also needs to be highlighted. Considering that communication link is not always available, the Delay Tolerant Networking (DTN) approach can be exploited, storing the data locally and conducting updates as soon as a network connection becomes available (Sawand et al. 2014).

The availability of e-health systems is critical to ensure the sensing and integrity of vital information collected from the patient. According to (Yang 2014), "*because resources, time, and money are always limited in real world applications, we have to find solutions to optimally use these valuable resources under various constraints.*" This is particularly true in the context of an e-health monitoring system. Historically, one of the barriers to the widespread adoption of telemedicine was cost factors associated with heterogeneous devices and single point solutions (often for therapeutic application) (Gatouillat et al. 2018). da Silva Lisboa et al. (2018) propose an e-health system architecture based on sensors, fog, and cloud computing. Stochastics models were proposed to evaluate how failures of the infrastructure impact the availability of the e-health system. The authors evaluated four different scenarios with different configurations. The experiments showed that the e-health system availability increases from 99.7299% to 99.9709% when the application present in the fog and in the cloud are complementary and has redundancy. In addition, sensitivity analysis results shown that the fog and cloud applications are the components that have higher impact on the e-health system application.

5.5 Conclusion

The Internet of Things is huge opportunity for both business and society. However, as we rely more and more on the Internet of Things, we require more availability and uptime, particularly where the criticality of the use case is high. This chapter discussed some of the challenges and solutions to address availability in smart cities, smart agriculture and the Internet of Medical Things.

Improving the availability of a system is not an easy task, especially if availability, including connectivity and coverage, are not considered in the conception phase. Some techniques, such as redundancy and geographic distribution, are commonly applied but at same time, from a management perspective, they bring many challenges.

When adopting resource redundancy and distribution, the service provider has to deal with both implementation and maintenance costs. Ibrahim et al. (2018) and da Silva Lisboa et al. (2018) demonstrate the availability gain of having a fault-tolerant approach with redundancy but they do not provide the cost analysis. Both physical and applications layers should be taken into account when analysing the costs of the system implementation and maintenance. The physical devices, connected through a network, host a set of software that most of the time is running on virtual machines (VMs). Orchestrating all these (physical and virtual) components is complex and demand specialist personnel or automated systems.

If the system is designed to scale (up and down), better resource utilization can be achieved and the system will be able to handle different workloads with different requirements. As a result, downtime due to poor scaling capacity will be reduced. Virtualization is a technique that supports service scalability by allowing the deployment of new service instances on demand. Similarly, machine learning approaches are being used to predict workloads and propose better resource planning (Liu et al. 2017; Le Duc et al. 2019). However, there is no way to avoid failures. When it happens, automatic failover approaches should be in place to minimize the system unavailability and reduce mean time to repair.

References

Ali, Samr, and Mohammed Ghazal. 2017. *Real-Time Heart Attack Mobile Detection Service (RHAMDS): An IoT Use Case for Software Defined Networks.* 2017 IEEE 30th Canadian Conference on Electrical and Computer Engineering (CCECE), 1–6. IEEE.

Aloqaily, Moayad, Ismaeel Al Ridhawi, Burak Kantraci, and Hussein T. Mouftah. 2017. *Vehicle as a Resource for Continuous Service Availability in Smart Cities.* 2017 IEEE 28th Annual International Symposium on Personal, Indoor, and Mobile Radio Communications (PIMRC), 1–6. IEEE.

Arasteh, H., V. Hosseinnezhad, V. Loia, A. Tommasetti, O. Troisi, M. Shafie-Khah, and P. Siano. 2016. *Iot-based Smart Cities: A Survey*. 2016 IEEE 16th International Conference on Environment and Electrical Engineering (EEEIC), 1–6. IEEE.

Araujo, Jean, Bruno Silva, Danilo Oliveira, and Paulo Maciel. 2014. *Dependability Evaluation of a Mhealth System Using a Mobile Cloud Infrastructure*. 2014 IEEE International Conference on Systems, Man, and Cybernetics (SMC), 1348–1353. IEEE.

Arroub, Ayoub, Bassma Zahi, Essaid Sabir, and Mohamed Sadik. 2016. *A Literature Review on Smart Cities: Paradigms, Opportunities and Open Problems*. 2016 International Conference on Wireless Networks and Mobile Communications (WINCOM), 180–186. IEEE.

Baldoni, Roberto, Luca Montanari, and Marco Rizzuto. 2015. On-line Failure Prediction in Safety-Critical Systems. *Future Generation Computer Systems* 45: 123–132.

Bawany, Narmeen Zakaria, and Jawwad A. Shamsi. 2015. Smart City Architecture: Vision and Challenges. *International Journal of Advanced Computer Science and Applications* 6 (11): 246–255.

Bu, Fanyu, and Xin Wang. 2019. A Smart Agriculture IoT System based on Deep Reinforcement Learning. *Future Generation Computer Systems* 99: 500–507.

Chakrabarty, Shaibal, and Daniel W. Engels. 2016. *A Secure IoT Architecture for Smart Cities*. 2016 13th IEEE Annual Consumer Communications & Networking Conference (CCNC), 812–813. IEEE.

Chiuchisan, Iuliana, Hariton-Nicolae Costin, and Oana Geman. 2014. *Adopting the Internet of Things Technologies in Health Care Systems*. 2014 International Conference and Exposition on Electrical and Power Engineering (EPE), 532–535. IEEE.

Darkins, Adam, Patricia Ryan, Rita Kobb, Linda Foster, Ellen Edmonson, Bonnie Wakefield, and Anne E. Lancaster. 2008. Care Coordination/Home Telehealth: The Systematic Implementation of Health Informatics, Home Telehealth, and Disease Management to Support the Care of Veteran Patients with Chronic Conditions. *Telemedicine and e-Health* 14 (10): 1118–1126.

Deloitte. 2018. *2018 Global Health Care Outlook—The Evolution of Smart Health Care*. Technical Report.

Le Duc, Thang, Rafael García Leiva, Paolo Casari, and Per-Olov Östberg. 2019. Machine Learning Methods for Reliable Resource Provisioning in Edge-Cloud Computing: A Survey. *ACM Computing Surveys (CSUR)* 52 (5): 94.

Farahani, Bahar, Farshad Firouzi, Victor Chang, Mustafa Badaroglu, Nicholas Constant, and Kunal Mankodiya. 2018. Towards Fog-Driven IoT e-Health: Promises and Challenges of IoT in Medicine and Healthcare. *Future Generation Computer Systems* 78: 659–676.

Gatouillat, Arthur, Youakim Badr, Bertrand Massot, and Ervin Sejdić. 2018. Internet of Medical Things: A Review of Recent Contributions Dealing with Cyber-Physical Systems in Medicine. *IEEE Internet of Things Journal* 5 (5): 3810–3822.

Gaur, Aditya, Bryan Scotney, Gerard Parr, and Sally McClean. 2015. Smart City Architecture and Its Applications based on IoT. *Procedia Computer Science* 52: 1089–1094.

Gondchawar, Nikesh, and R.S. Kawitkar. 2016. IoT based Smart Agriculture. *International Journal of Advanced Research in Computer and Communication Engineering* 5 (6): 838–842.

Grand View Research. 2019. Smart Cities Market Analysis Report by Application (Governance, Buildings, Utilities, Transportation, Healthcare, Environmental Solution), By Region, And Segment Forecasts, 2019–2025.

He, Jianhua, Jian Wei, Kai Chen, Zuoyin Tang, Yi Zhou, and Yan Zhang. 2017. Multitier Fog Computing with Large-Scale IoT Data Analytics for Smart Cities. *IEEE Internet of Things Journal* 5 (2): 677–686.

Ibrahim, Hassan, Norhan Mostafa, Hassan Halawa, Malak Elsalamouny, Ramez Daoud, Hassanein Amer, Amr Shaarawi, Ahmed Khattab, and Hany ElSayed. 2018. *A High Availability Networked Control System Architecture for Precision Agriculture*. International Conference on Computer and Applications (ICCA), 457–460. IEEE.

Kakderi, Christina, Nicos Komninos, and Panagiotis Tsarchopoulos. 2019. Smart Cities and Cloud Computing: Lessons from the STORM CLOUDS Experiment. *Journal of Smart Cities* 1 (2): 4–13.

Liu, Ning, Zhe Li, Jielong Xu, Zhiyuan Xu, Sheng Lin, Qinru Qiu, Jian Tang, and Yanzhi Wang. 2017. *A Hierarchical Framework of Cloud Resource Allocation and Power Management Using Deep Reinforcement Learning*. 2017 IEEE 37th International Conference on Distributed Computing Systems (ICDCS), 372–382. IEEE.

McKinsey. 2016. How Big Data will Revolutionize the Global Food Chain. https://www.mckinsey.com/business-functions/mckinsey-digital/our-insights/how-big-data-will-revolutionize-the-global-food-chain#

Mugarza, Imanol, Andoni Amurrio, Ekain Azketa, and Eduardo Jacob. 2019. Dynamic Software Updates to Enhance Security and Privacy in High Availability Energy Management Applications in Smart Cities. *IEEE Access* 7: 42269–42279.

Mulla, David J. 2013. Twenty Five Years of Remote Sensing in Precision Agriculture: Key Advances and Remaining Knowledge Gaps. *Biosystems Engineering* 114 (4): 358–371.

Ossebaard, Hans Cornelis, A.C.P. De Bruijn, Julia E.W.C. van Gemert-Pijnen, and R.E. Geertsma. 2013. Risks Related to the Use of e-Health Technologies: An Exploratory Study.

Rahmani, Amir M., Tuan Nguyen Gia, Behailu Negash, Arman Anzanpour, Iman Azimi, Mingzhe Jiang, Pasi Liljeberg. 2018. Exploiting smart e-Health gateways at the edge of healthcare Internet-of-Things: A fog computing approach. *Future Generation Computer Systems* (78): 641–658.

Santos, Guto Leoni, Patricia Takako Endo, Matheus Felipe Ferreira da Silva Lisboa, Leylane Graziele Ferreira da Silva, Djamel Sadok, Judith Kelner, and Theo Lynn. 2018. Analyzing the Availability and Performance of an e-Health System Integrated with Edge, Fog and Cloud Infrastructures. *Journal of Cloud Computing* 7 (1): 16.

Sawand, Ajmal, Soufiene Djahel, Zonghua Zhang, and Farid Naït-Abdesselam. 2014. *Multidisciplinary Approaches to Achieving Efficient and Trustworthy eHealth Monitoring Systems*. 2014 IEEE/CIC International Conference on Communications in China (ICCC), 187–192. IEEE.

da Silva Lisboa, Matheus Felipe Ferreira, Guto Leoni Santos, Theo Lynn, Djamel Sadok, Judith Kelner, and Patricia Takako Endo. 2018. *Modeling the Availability of an e-Health System Integrated with Edge, Fog and Cloud Infrastructures*. 2018 IEEE Symposium on Computers and Communications (ISCC), 00416–00421. IEEE.

Soyturk, Mujdat, Khaza N. Muhammad, Muhammed N. Avcil, Burak Kantarci, and Jeanna Matthews. 2016. *From Vehicular Networks to Vehicular Clouds in Smart Cities*. Smart Cities and Homes, 149–171. Morgan Kaufmann.

Stantchev, Vladimir, Ahmed Barnawi, Sarfaraz Ghulam, Johannes Schubert, and Gerrit Tamm. 2015. Smart Items, Fog and Cloud Computing as Enablers of Servitization in Healthcare. *Sensors & Transducers* 185 (2): 121.

Tachakra, Sapal, X.H. Wang, Robert S.H. Istepanian, and Y.H. Song. 2003. Mobile e-Health: The Unwired Evolution of Telemedicine. *Telemedicine Journal and e-Health* 9 (3): 247–257.

Tang, Dong, Dileep Kumar, Sreeram Duvur, and Oystein Torbjornsen. 2004. *Availability Measurement and Modeling for an Application Server*. International Conference on Dependable Systems and Networks, 2004, 669–678. IEEE.

Tang, Bo, Zhen Chen, Gerald Hefferman, Tao Wei, Haibo He, and Qing Yang. 2015. *A Hierarchical Distributed Fog Computing Architecture for Big Data Analysis in Smart Cities*. Proceedings of the ASE BigData & SocialInformatics 2015, 28. ACM.

TongKe, Fan. 2013. Smart Agriculture based on Cloud Computing and IOT. *Journal of Convergence Information Technology* 8 (2).

Topol, Eric. 2015. *The Patient will See You Now: The Future of Medicine is in Your Hands*. Basic Books.

Vargheese, Rajesh, and Yannis Viniotis. 2014. *Influencing Data Availability in IoT Enabled Cloud based e-Health in a 30 Day Readmission Context*. 10th IEEE International Conference on Collaborative Computing: Networking, Applications and Worksharing, 475–480. IEEE.

Yang, Xin-She. 2014. *Nature-Inspired Optimization Algorithms*. Elsevier.

Zamora-Izquierdo, Miguel A., José Santa, Juan A. Martínez, Vicente Martínez, and Antonio F. Skarmeta. 2019. Smart Farming IoT Platform based on Edge and Cloud Computing. *Biosystems Engineering* 177: 4–17.

CHAPTER 6

Security of Distributed Intelligence in Edge Computing: Threats and Countermeasures

Mohammad S. Ansari, Saeed H. Alsamhi, Yuansong Qiao, Yuhang Ye, and Brian Lee

Abstract Rapid growth in the amount of data produced by IoT sensors and devices has led to the advent of edge computing wherein the data is processed at a point at or near to its origin. This facilitates lower latency, as well as data security and privacy by keeping the data localized to the edge node. However, due to the issues of resource-constrained hardware and software heterogeneities, most edge computing systems are prone to a large variety of attacks. Furthermore, the recent trend of incorporating intelligence in edge computing systems has led to its own security issues such as data and model poisoning, and evasion attacks. This chapter presents a discussion on the most pertinent threats to edge intelligence.

M. S. Ansari (✉) • Y. Qiao • Y. Ye • B. Lee
Software Research Institute, Athlone IT, Athlone, Ireland
e-mail: mansari@ait.ie; ysqiao@research.ait.ie; yye@research.ait.ie; blee@ait.ie

S. H. Alsamhi
IBB University, Ibb, Yemen
e-mail: saeedalsamhi@gmail.com

T. Lynn et al. (eds.), *The Cloud-to-Thing Continuum*, Palgrave
Studies in Digital Business & Enabling Technologies,
https://doi.org/10.1007/978-3-030-41110-7_6

Countermeasures to deal with the threats are then discussed. Lastly, avenues for future research are highlighted.

Keywords Edge AI • Edge computing • Distributed intelligence • Federated learning • Threats to Edge AI

6.1   EDGE COMPUTING: THREATS AND CHALLENGES

As discussed in Chap. 1, edge computing refers to data processing at or near the point of its origin rather than onward transmission to the fog or cloud. The 'edge' is defined as the network layer encompassing the smart end devices and their users, and is identified by the exclusion of cloud and fog (Iorga et al. 2018). For instance, a smartphone is the edge between body things and the cloud, and a gateway in a smart home is the edge between home things and the cloud (Shi et al. 2016).

Although edge computing brings a lot of advantages, and is being used in a variety of scenarios, it is not without its share of security threats and challenges. In fact, the following factors work towards expanding the attack surface in the case of edge computing:

Hardware Constraints: Since most edge computing hardware (edge devices, and even edge servers) have lower computational power and storage capacity as compared to a fog or cloud server, they are incapable of running dedicated attack prevention systems like firewalls, and are therefore more vulnerable to attacks.

Software Heterogeneities: Most devices and servers operating in the edge layer communicate using a large variety of protocols and operating systems without a standardized regulation. This makes the task of designing a unified protection mechanism difficult.

Most of these threats are exacerbated due to design flaws, implementation bugs, and device misconfigurations in the edge devices and servers (Xiao et al. 2019). Also, the lack of full-fledged user interfaces in many edge devices often makes it impossible to discern an ongoing/transpired attack.

In light of the above, understanding the security threats (and defenses) in edge computing assumes utmost importance. This section presents an overview of the state-of-the-art in the security threats and countermeasures employed in edge computing.

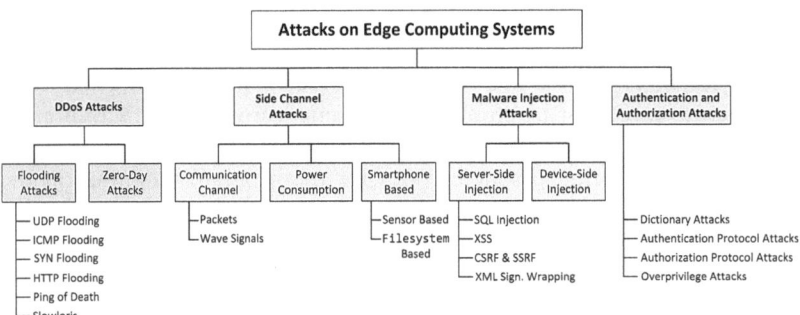

Fig. 6.1 Different types of attacks against edge computing systems (Xiao et al. 2019)

As depicted in Fig. 6.1, most attacks on edge computing infrastructure may be placed in one of the following four categories: DDoS attacks, side-channel attacks, malware injection attacks, and authentication and authorization attacks (Xiao et al. 2019). Each of these attacks and the countermeasure devised to deal with the corresponding attacks are discussed next.

6.1.1 DDoS Attack

In this type of attack, the goal of the adversary is to engage all the resources and bandwidth available at the target in order to prevent legitimate users from using the victimized system. In a typical DDoS attack, the attacker persistently sends a huge number of packets to the target (also referred to as 'flooding') thereby ensuring that all the resources of the target are exhausted in handling the malicious packets, and therefore genuine requests cannot be processed. Such attacks assume greater importance in the edge computing paradigms as they are computationally less powerful (than cloud servers), and therefore cannot run strong defense systems. Such attacks may be further categorized as UDP flooding attacks, ICMP flooding, SYN flooding, ping of death (PoD), HTTP flooding, and Slowloris (Xiao et al. 2019). Apart from the flooding attacks, another type of DDoS attack is a zero-day attack in which an attacker finds and utilizes a still-unidentified vulnerability in the target system to cause system shutdown.

Defenses and Countermeasures. Most potent solutions against flooding attacks utilize the detect-and-filter technique. The detection of malicious flooding packets may either be on a per-packet basis wherein each individual packet is inspected and discarded if deemed to be suspicious, or on a statistical basis wherein malicious packets are identified using parameters like packet entropy or by employing machine learning tools. Countering zero-day attacks on edge computing hardware is more difficult due to the unavailability of original source codes for the programs running on the machine, and also due to the fact that in many cases the software comes embedded in a firmware and is not amenable for inspection.

6.1.2   Side-Channel Attacks

These attacks operate by first capturing publicly available, non-privacy-sensitive information pertaining to the target (also called the side-channel information), and then inferring the private and protected data from this information by exploiting the correlations that are inherently present between the public and the private information. Typical examples of such attacks include capturing communication signals (e.g. packets or wave signals) to leak user's private data, monitoring the power consumption of edge devices to reveal usage patterns, and targeting the filesystem (e.g. the /proc filesystem in Android) and sensors (e.g. microphone, camera) on end devices like smartphones.

Defenses and Countermeasures: Due to their passive nature, side-channel attacks are difficult to defend against. Some commonly suggested defense mechanisms include data perturbation and differential privacy. The most popular data perturbation algorithm is *k*-anonymity which modifies the identifier information in the data prior to publishing its sensitive attributes. Lastly, it is important to note that ironically most defense mechanisms are themselves vulnerable to side-channel attacks (Xiao et al. 2019).

6.1.3   Malware Injection Attacks

The infeasibility of installing a full-fledged firewall on resource-constrained edge devices makes them vulnerable to malware injection attacks, wherein an attacker stealthily installs malicious programs in a target system. Such malware injection may either be performed at the edge server or the edge device(s). Server-side injection attacks can further be divided into four types: SQL injection, cross-site scripting (XSS), XML signature wrapping, and Cross-Site Request Forgery (CSRF) & Server-Site Request Forgery

(SSRF) (Xiao et al. 2019). Device-side injection attacks typically target the firmware of the end devices.

In a SQL injection attack, the attacker aims to destroy the backend database by sending carefully crafted SQL queries containing malicious executable codes. In a XSS attack, the adversary injects malignant HTML/JavaScript codes into the data content which may be accessed and executed by a server leading to its compromise. A CSRF attack is one in which the edge server is tricked into executing malicious programs embedded in web applications, and a SSRF attack is carried out by compromising and using an edge server to alter the internal data and/or services. Lastly, an XML signature wrapping attack works by intercepting and modifying a XML message, and re-transmitting it to a target machine in order to run tainted code.

Defenses and Countermeasures: To counter the server-side injection attacks, the detect-and-filter technique has been shown to be the most promising. Defense mechanisms against injection attacks generally rely on static analysis for malicious code detection and fine-grained access control. Research on devising means to mitigate firmware modification is also being carried out for prevention of such attacks.

6.1.4 Authentication and Authorization Attacks

The authentication and authorization processes in edge computing systems may also be susceptible to attacks. Such attacks may be put into four different categories: dictionary attacks, attacks targeting vulnerabilities in authentication mechanisms, attacks exploiting susceptibilities in authorization protocols, and over-privileged attacks (Xiao et al. 2019). Dictionary attacks employ a credential/password dictionary to get past the authentication systems. Attacks targeting vulnerabilities in authentication mostly work by utilizing loopholes in the WPA/WPA2 security protocols. Authorization based attacks exploit the logical weaknesses or design flaws that may exist in authorization protocols used by the edge computing systems. In over-privileged attacks, the attacker tricks the victim system into assigning higher (than required) access rights to an app or device, which can then be used to perform malicious activities inside the network.

Defenses and Countermeasures: The most potent defense against dictionary attacks is the addition of one more layer of authentication (typically known as two-factor authentication). To counter the attacks which target authentication protocols, two common approaches are enhancing the security of the communication protocols, and hardening the crypto-graphic implementation. The OAuth 2.0 protocol is the best defense against authorization attacks, and has been proven to be theoretically secure. To counter the over-privileged attacks, the most effective solution involves strengthening the permission models for the operating systems running on edge devices.

Most of the security threats and challenges, along with the associated countermeasures, discussed above pertain to edge computing systems which are configured as passive data aggregation and processing nodes with little to no intelligence built into them. However, the recent trend of incorporation of intelligence (in the form of inference generation, and even on-device training, in the context of machine learning) into the edge nodes/devices, brings its own share of issues and challenges, and the need for specialized defenses and countermeasures.

This chapter aims to highlight the threat landscape for the scenario where edge devices are becoming smarter with the inclusion of machine learning. Therefore, the remainder of the chapter focuses on the tech-niques for incorporation of intelligence into edge computing systems, the security threats associated with such systems, and the pertinent counter-measures and defenses that have been devised against attacks on edge intelligence. Section 6.2 presents a discussion on the need for, and the techniques to bring intelligence to the edge computing systems. Security threats targeted towards intelligent edge systems are highlighted in Sect. 6.3 (For a quick summary, please refer to Table 6.1). Techniques that have been developed to defend against the threats, and mitigate the attacks on edge computing systems are discussed in Sect. 6.4. Section 6.5 contains a discussion on future research directions in the field of intelligent edge computing. Section 6.6 presents concluding remarks.

6.2   EDGE INTELLIGENCE

The incorporation of artificial intelligence into the constituents of edge layer is referred to as *Edge AI*. The two biggest advantages of Edge AI are briefly discussed below.

Faster Inference: For applications which utilize a pre-trained machine learning model to output classifications or predictions, processing data at the edge leads to faster results. This is primarily due to the elimina-tion of the data transfer time between the edge and the cloud.

Table 6.1 Security threats to edge computing systems, defense mechanisms, and assets targeted by the different attacks

Attack	Type	Sub-type	Ref.	Defense	Asset targeted
DDoS attack	Flooding based	UDP flooding ICMP flooding	Xiaoming et al. (2010) Udhayan and Anitha (2009)	Detect-and-filter • Per packet based detection	• Network infrastructure • Virtualization infrastructure
		SYN flooding HTTP flooding	Bogdanoski et al. (2013) Dhanapal and Nithyanandam (2017) Sonar and Upadhyay (2014)	• Statistics-based detection	
		Ping of death			
		Slowloris	Damon et al. (2012)		
	Zero-day		NIST (2010)	Code-level vulnerability identification	
Side-channel attack	Exploit communication channels	Packets Wave signals	Chen and Qian (2018) Enev et al. (2011)	Data perturbation • k-anonymity • Differential privacy	• User data • User privacy
	Exploiting power consumption data		Örs et al. (2003)	Restricting access to side-channels	
	Target smart devices	OS based Sensor based	Zhou et al. (2013) Chen et al. (2018b)		
Malware injection attacks	Server-side injection	SQL injection XSS CSRF & SSRF XML signature wrapping	Anley (2002) Cisco (2016) Costin (2018) McIntosh and Austel (2005)	Detect-and-filter	• Edge server • Edge devices
	Device-side injection	Buffer overflow	Greenberg (2017)	Code-level analysis	

(continued)

Table 6.1 (continued)

Attack	Type	Sub-type	Ref.	Defense	Asset targeted
Authentication and authorization attacks	Dictionary attack		Nakhila (2015)	Two-factor authentication	• Edge server • Virtualization infrastructure • Edge devices
	Authentication protocol attack		Vanhoef and Piessens (2018)	Hardening authentication protocols • Enhance security of protocol • Secure cryptographic implementation	
	Authorization protocol attack		Chen (2014)	Hardening authorization protocols • OAuth 2.0	
	Over-privileged attacks		Sun and Beznosov (2012)	Strengthening permission models for mobile OS	

Data Locality: Since most of the data processing and inference is performed it the edge layer, the data actually never leaves this layer (and is not sent to the fog/cloud). Such data locality is of paramount importance in safeguarding user privacy in applications like health monitoring, indoor localization, etc. Further, keeping the data on or near the source, and not transferring it to the cloud (which may be in a different country), alleviates regulatory/legal issues pertaining to the data.

Although the advantage of faster inference with the data remaining localized is interesting, the resource constraints in most constituents of the edge layer dictate that specialized techniques have to be employed for performing inference and training in Edge AI.

6.2.1 Lightweight Models for Edge AI

The first case is where an edge computing node is only used for inference using a pre-trained model. In such cases, the emphasis is to build lightweight models capable of running in resource constrained environments. This discussion will focus on image processing models because a major portion of available research on light models for Edge AI deals with computer vision. This is driven by the success of Convolutional Neural Networks (CNN) for image recognition and classification tasks, albeit with huge computational requirements. AlexNet was the first CNN variant which employed a technique called Group Convolution to reduce the number of parameters, and resulted in a 240 MB sized model (Krizhevsky et al. 2012). Xception used a more stringent version of group convolution to further reduce the number of model parameters (88 MB model size) (Chollet 2017). GoogleNet managed to reduce the parameter size to 27 MB while maintaining the accuracy (Szegedy et al. 2015). However, the breakthrough which enabled CNN variants to be used on edge devices was MobileNet (Howard et al. 2017), which required approximately 8–9 times less computation than standard CNN, and had model size of 16 MB (Howard et al. 2017). MobileNet*V2* further provided a performance improvement while reducing the model size to 14 MB (Sandler et al. 2018). SqueezeNet is even more efficient, and is capable of providing AlexNet level accuracy with only 5 MB of parameters (Iandola et al. 2016), which is a sufficiently small sized model for deployment on low-complexity embedded hardware like Raspberry Pi.

6.2.2   Data and Model Parallelism

For cases where the edge computing nodes are to be used for training as well, techniques like data parallelism and model parallelism are employed.

Data Parallelism: In data parallelism, the training dataset is divided into non-overlapping partitions and fed to the participating nodes. Figure 6.2(a) depicts the data parallelism applied to a group of three machines. All nodes train the complete model using a subset of data. The advantage is that the training task is performed at multiple nodes concurrently (for different data sub-sets). Specialized algorithms like Synchronous Stochastic Gradient Descent (Sync-SGD) (Das et al. 2016), and Asynchronous Stochastic Gradient Descent (Async-SGD) (Zhang et al. 2013) have been devised to ensure timely and efficient update of the global weights and parameters of the model.

Model Parallelism: In model parallelism, the ML model is divided into partitions and each participant node is responsible for maintaining one partition. Figure 6.2(b) depicts the model parallelism applied to a group of four machines. Designing the model partitions is non-trivial and NP-complete in this case, as the participating machines may have different storage, computing, and networking capabilities (Dean et al. 2012). Further, dividing the training dataset is also not straightforward in this case, as the logical partitions have to be decided in accordance with the partition scheme of the input layer.

To reduce the communication of a large number of parameters between participating devices, *model compression* is used. It has been demonstrated that quantizing the parameter bitwidth from 32 bits to 8 bits does not

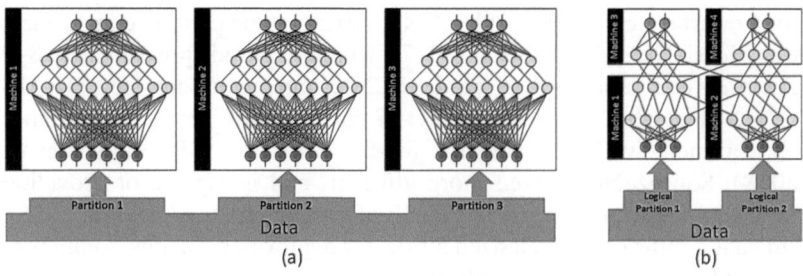

Fig. 6.2   (a) Data parallelism and (b) model parallelism

impact the accuracy of CNN-like architectures significantly (Cheng et al. 2017). Further, reducing the communication overhead by quantizing the gradients (computed using Stochastic Gradient Descent) is referred to as *Gradient Compression* or *Gradient Quantization*.

6.2.3   Federated Learning

Data collected by a lot of devices may not be amenable for sharing over a cloud due to reasons of privacy. Examples include data collected by health monitoring devices, CCTV recordings, etc. For such cases, a distributed ML technique called Federated Learning (FL) has been proposed (Konečný et al. 2016), which enables smart devices to collaboratively learn a shared prediction model while keeping all the training data on device. This effectively decouples the learning process from the need to store the data centrally, and goes beyond the use of pre-trained models to make predictions on mobile devices by bringing model training to the device. As shown in Fig. 6.3, FL works by first downloading the current model to an edge device. Thereafter, the model is updated using locally stored data, and updates are forwarded to a central server where they undergo a

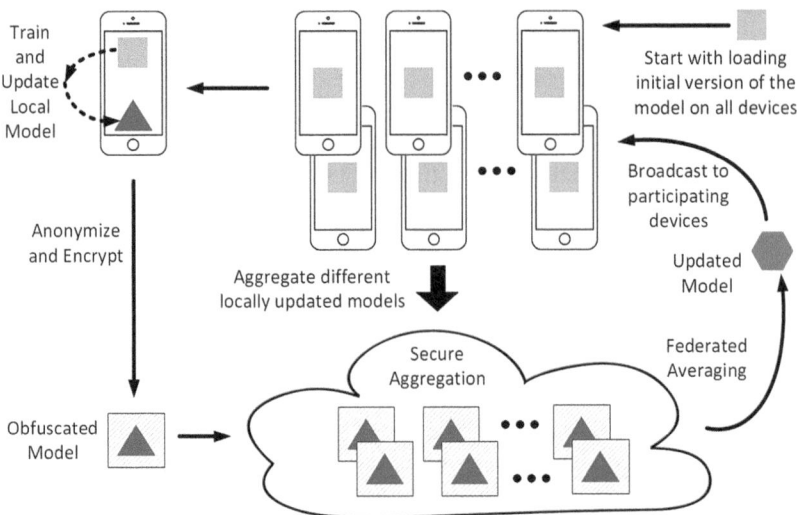

Fig. 6.3   Federated learning over multiple smartphones

Federated Averaging with the updates from other users. Since the user data never leaves the device, and individual updates are not stored in the cloud, data security and privacy is ensured.

The updates in this case are not simple gradient updates as in the case of conventional distributed ML models. Instead, high-quality updates containing much more information than just the changes in gradients are computed, compressed, and sent for processing. This ensures convergence with reduced communication (up to 100 times (Konečný et al. 2016)) between the edge device and the central server. Scheduling algorithms are used to ensure that training happens only when the device is idle, charging, and on a free wireless connection, so there is no degradation in the end-user experience. With most flagship phones nowadays coming with a dedicated AI chip, there are estimated to be approximately two billion smartphones with underutilized processing capability. Federated Learning can leverage this enormous pool of computing resources to improve existing models, or to train new ones from scratch.

The distribution of intelligence over a multitude of end devices is therefore slated to bring significant improvements in the way conventional IoT devices function. However, this distribution of intelligence to the edge nodes also opens up a plethora of security issues which are discussed next.

6.3   THREATS TO EDGE AI

Despite their widespread usage by virtue of the advantages they offer, Edge AI paradigms are not without their share of limitations and points of concerns. Incorporating intelligence in the edge layer is a double edged sword in the sense that although the impact of a potential attack is limited to a localized environment, the less potent security protocols on the resource-constrained edge hardware make them more vulnerable to attacks. The situation is further aggravated by the casual attitude of human operators responsible for the configuration and maintenance of the edge devices. For instance, a survey of 439 million households using WiFi networks showed that approximately 50% of them were unsecured, and of the remaining, 80% have their router still configured with the default passwords (Shi et al. 2016). The figure is even poorer for public WiFi hotspots, with 89% of them being unsecured or poorly configured (Shi et al. 2016). Furthermore, updating or re-configuration of the security software on edge devices is non-trivial because there may be legacy devices for which support has ended, or the constrained hardware resources available on the device may present restrictions on the authentication protocols that could be run on the device. Moreover, the heterogeneous nature of the edge networks means that there can be no uniform security policy. Lastly, microservers used in the edge computing environment lack the hardware protection mechanisms available on commodity servers (Roman et al. 2018).

A discussion on the threats to Edge AI systems can be divided into two distinct cases: threats to Edge AI used for inference, and threats to Edge AI used for learning/training (as in Federated Learning). Each of these scenarios are discussed separately below. It needs to be mentioned that in the discussion that follows, it is considered that the intelligence is located in the edge device. However, this is not a restrictive scenario. In fact, the attacks and countermeasure discussed below are equally relevant in the case where the machine intelligence is located in an edge server or a gateway.

6.3.1   Threats to Edge AI for Inference

The vast majority of Edge AI deployments at present are used for inferencing based on pre-trained models. This is suitable for edge devices due to the limited computing resources they offer. As discussed before, there has been progress in model compression that allow high performance models (e.g. SqueezeNet) to be run in resource-constrained environments. In such a standalone environment, where the edge devices use the pre-trained model independently, the most probable attack is the feeding of adversarial examples to the model thereby causing the model to output incorrect predictions. Such attacks are referred to as Evasion Attacks and are discussed next.

6.3.2   Evasion Attacks

The susceptibility of machine learning models to adversarial samples, which essentially are carefully perturbed inputs that look and feel exactly the same as their untampered counterparts to a human, is well documented (Biggio and Roli 2018). Although it may seem that adversarial examples are available only for image recognition models (Kurakin et al. 2016), the earliest reported instance of such an attack is for a machine learning based email spam filter, wherein it was shown that linear classifiers could be tricked easily by carefully crafted changes in the text (Dalvi et al.

2004). It is still not proven why adversarial samples work, but a commonly accepted hypothesis, called the tilted boundary hypothesis, asserts that since the model can never fit the data perfectly (at least theoretically), there would always be adversarial pockets of inputs existing between the classifier boundary and the actual sub-manifold of sampled data (Szegedy et al. 2013). Since the models devised to be used in the low resource environments of edge computing are compressed variants of bigger, deeper, and more robust models, these are generally more prone to such adversarial attacks. Evasion attacks can be of different types: Gradient based, Confidence-score based, Hard Label based, Surrogate model based, and Brute-force attacks (Moisejevs 2019).

Gradient based attacks require access to the model gradients (and thus belong to the category of Whitebox attacks). Theoretically, such attacks are the most potent as the attacker may use the model gradients to gain insights into the working of the model, and can then mathematically optimize the attack. This approach is the most probable one to target hardened models, as it has been shown that if an adversary has access to the model gradients, it is *always* possible to generate adversarial samples irrespective of the robustness of the model (Carlini and Wagner 2017). Some examples of such attacks include Elastic-Net attack based on L_1 norm (Chen et al. 2018a), an L_2 Norm based attack (Carlini and Wagner 2017), and an L_∞ Norm based attack (Madry et al. 2017).

Confidence-score based adversarial attacks utilize the output confidence score to get estimates of the gradients of the model. The adversary may then use these estimated gradients to orchestrate an attack similar to the gradient based attack. Since this approach does not require any information about the composition of the model, this attack may be classified as a Blackbox attack. Examples include the Zeroth Order Optimization based attack (Chen et al. 2017a), the Natural Evolutionary Strategies (NES) based attack (Ilyas et al. 2018), and the Simultaneous Perturbation Stochastic Approximation (SPSA) based attack (Uesato et al. 2018).

Label based attacks rely on estimating the gradients by using the hard labels generated by the model. Since only the label information is required by the adversary, such attacks are generally simple to implement, and require little hyperparameter tuning. Boundary Attack is the most powerful attack in this category. It works by starting from a large adversarial perturbation and seeks to incrementally reduce the perturbation while staying adversarial (Brendel et al. 2017).

Surrogate model based attacks first try to build a replica of the target model. If the internals of the target model are not known, the adversary can reverse engineer the structure of the model by repeatedly querying the target model and observing the input-output pairs. If the target model is not available for querying, then the attacker can start by guessing the architecture in the case of the model being applied for a standard machine learning problem like image classification (Moisejevs 2019). Thereafter, the gradient based attack can be fine-tuned on this surrogate model, and then used on the actual model.

Lastly, Brute-force attacks, as the name implies, work by generating adversarial examples by resorting to transformations, perturbations and addition of noise to the data samples. Such attacks do not rely on mathematical optimization, and therefore require no knowledge of the model. Such an approach is generally used by adversaries who have access to large computational resources, and do not have a timeline for the success of their attacks.

6.3.3 Privacy Attacks

The previous section discussed the issues pertaining to evasion attacks wherein the goal of the attacker is to cause the model to output incorrect predictions. However, there is another class of attacks, known as Privacy Attacks, which aim to siphon off valuable information from the data used by the model. For instance, an adversary may be interested in knowing whether a certain person is enrolled in a healthcare program. There are several other examples of such private information which an attacker may want to unravel: credit card details, location information, and household energy consumption. While the risk with disclosure of credit card information is obvious, the availability of location and energy usage information of a person can inform the attacker about when the person is away for a vacation (consequently leaving his house unattended). There are two broad categories of such privacy attacks on machine learning systems:

Membership Inference Attacks: This is the case when the adversary has one or more data points, and wants to ascertain whether the data points were part of the training set or not (Shokri et al. 2017). For instance, an attacker might want to find out whether a given person X is included in a critical illness list in the healthcare records of a state. Such attacks are increasingly being targeted towards recommender systems, wherein the training dataset may contain information such as gender, age, ethnicity, location, sexual orientation, immigration status, political affiliation, net worth and buying preferences. An attacker who knows a few pieces of information from these may be able to expose other details using membership inference. A detailed study of such attacks has been carried out (Truex et al. 2019), which concluded that several factors affected the potency of membership inference attacks. Firstly, the model becomes more vulnerable with increase in the number of classes. Also, the choice of the algorithm for training is also an important factor. Algorithms whose decision boundaries are not significantly impacted by an individual training sample are less vulnerable.

Model Inversion Attacks: Such attacks, also known as Data Extraction attacks, work by extracting an average representation of each of the classes the target model was trained on. For instance, a model trained for facial recognition may be attacked in the following manner. First, a base image is chosen based upon the physical characteristics (age, gender, ethnicity) of the person whose image is to be extracted from the model. Then the attacker can repeatedly query the target model with different modifications in the base image, until a desired confidence level is reached. It has been shown that the final image in such an attack scenario can be fairly demonstrative of the face of the person concerned (Fredrikson et al. 2015). With the increasing integration of ML based face recognition systems in modern day security and surveillance setups including the ones at airports, such attacks may lead to the divulgence of private and sensitive information like photographs, visa and passport details, travel itineraries, and much more. In another instance, it has been demonstrated that it is possible to extract credit card details and social security numbers from a text generator trained on private data (Carlini et al. 2019).

6.3.4   Threats to Edge AI for Training

This section deals with the threats that are pertinent for Edge AI systems which are used for performing both machine learning training and inference. Firstly, the convergence guarantee of the federated learning algorithms has not still been theoretically established (Ma et al. 2019). Only approximate convergence may be guaranteed, and that too requires some unrealistic assumptions: (1) training data is shared across devices or distributed amongst the participating devices in an independent and identically distributed (IID) manner, and (2) all participating devices are involved in communication of updates for each round.

Secondly, in the federated learning scenario, an adversary can take control over one or more participating devices to inject spurious and arbitrary updates in order to manipulate the training process. This is generally referred to as *model poisoning* or *logic corruption*. Also, a malicious intruder may also compromise the training data in order to adversely affect the training process. This is commonly known as *data poisoning*, and may be in the form of either the manipulation of the labels in the training data, or the modification of the input itself. It has been shown that an adversarial participant can infer properties associated with a subset of training data (Bagdasaryan et al. 2018). Also, there may exist eavesdroppers on the broadcast link used by the centralized server to communicate the intermediate model state to the participants. Another way of classifying the poisoning attacks on Edge AI systems can be based on the characteristic that is targeted to be compromised. For instance, attacks targeting the *availability* of the system generally work by injecting a lot of spurious data into the training set, thereby ensuring that whatever classification boundary the model learns becomes useless. It has been shown that a 3% poisoning of the dataset can lead to more than 10% drop in accuracy (Steinhardt et al. 2017). Such attacks are the ML counterparts to the conventional Denial-of-Service attacks. Another class of attacks do not aim to affect the availability of the ML system, and instead target the *integrity* of the system. Such attacks are more sophisticated than availability attacks, and leave the classifier functioning exactly as it should, but with one or more backdoor inputs embedded into the model. These backdoor inputs cause the classifier to output incorrect predictions thereby compromising the integrity of the model. An example of such a backdoor input is a spam email checking scenario wherein an attacker teaches a model that if a certain string is present in the input, then that input is to be classified as benign (Chen et al. 2017b).

Further, although the concept of federated learning is appealing, it remains to be seen how it performs with scaling up. Several practical issues are expected to creep up when the FL systems are scaled up to involve a huge number of devices: limited device storage, unreliable connectivity, and interrupted execution. Moreover, it is still unknown whether a significant increase in the number of participating devices would translate to better accuracy and/or faster convergence of the model.

There can be another way of looking at the threats that may affect Edge AI. Typically, an Edge AI system is composed of three major components: network, services, and devices. The network (generally wireless network) may be susceptible to DoS and man-in-the-middle attacks, as well as prone to disruptions by a rogue node or gateway. The services running on the nodes may be infiltrated to cause privacy leakage, privilege escalation, and service manipulation. Lastly, the edge devices may themselves be prone to physical damage, as well as data poisoning.

6.4   COUNTERING THE THREATS TO EDGE AI

This section presents a discussion on the techniques available for dealing with the threats against Edge AI. Since the threats could be against the data, the model, or even the entire system (e.g. Federated Learning), the following discussion is structured accordingly. At the onset, it needs to be mentioned that no available countermeasure can be guaranteed to completely eliminate the threats to Edge AI systems, and it is by a judicious mix of the defense techniques that we can hope for a reasonable safe system.

6.4.1   Defenses against Data Poisoning

In a data poisoning attack on a machine learning system, the adversary injects malicious samples into the training pool. These tainted data samples are typically significantly different from the benign data points, and are therefore 'outliers.' The process of outlier detection (also known as anomaly detection or data sanitization) aims to identify and eliminate such outliers *before* the training process (Paudice et al. 2018). The anomaly detection process is obviously ineffective if the poisoned samples were introduced into the training dataset before the filtering rules were created. Further, if the attacker is able to generate data poison samples which are very similar to the pristine samples ('inliers'), then this line of defense breaks down. Another variant of the anomaly detection approach is the use of micromodels (Cretu et al. 2008). The Micromodel approach was first proposed for use in network intrusion detection datasets, wherein multiple micromodels were generated by training the classifier on non-overlapping slices of the training sets (micromodels of the training set). A majority voting scheme was then used on the micromodels to ascertain which of the training slices were corrupted by poisoning. The institution behind this approach is that network attacks are generally of a low time duration, and can only affect a few training slices.

Another commonly used defense technique is to analyze the effect of a new sample on the model's accuracy before actually including that sample in the training set. For a tainted data sample used as a test sample, the model's accuracy would degrade. Reject on Negative Impact (RONI) (Nelson et al. 2009), and target-aware RONI (tRONI) (Suciu et al. 2018) are defensive methods that use this approach. The RONI defense has been demonstrated to be extremely successful against dictionary attacks on email spam filters, identifying 100% of malicious emails without flagging any benign emails. However, RONI fails to mitigate targeted attacks because the poison instances in such cases might not individually cause a significant performance drop. Target-aware RONI was then proposed as a targeted variant which is capable of identifying instances that distort the target classification significantly.

A perturbation approach has also been employed for anomaly detection (Gao et al. 2019). STRong Intentional Perturbation (STRIP) intentionally perturbs the incoming data samples, for instance by superimposing different patterns on sample images, and observes the randomness of the predicted classes for the perturbed inputs. It is expected that a benign classifier would be affected significantly by the perturbations. A low entropy in the classes predicted by the model defies the input-dependence property of a pristine model and implies the presence of a tainted input.

Another method known as TRIM has been proposed for regression learning. It estimates the parameters iteratively, while employing a trimmed loss function to remove samples which lead to large residuals. It has been demonstrated that TRIM is able to isolate most of the poisoning points and learn a robust regression model (Jagielski et al. 2018).

Lastly, even after significant strides in automated anomaly detection, the role of human factors in identifying malicious data samples cannot be completely eliminated. Human-in-the-loop approach works by focusing the attention of human data analysts on outliers which cause an unwarranted boundary shift in a classifier model (Mei and Zhu 2015).

6.4.2 Countering Adversarial Attacks

Defenses against evasion attacks may be put into two broad categories: formal methods and empirical approaches. Formal methods are purely mathematical in nature, and work by testing the model on all possible adversarial samples which can be generated within the allowable limits of perturbation. While this approach leads to virtually impenetrable models, the method is not amenable to most present day applications of machine learning due to its high requirement of computational resources. For instance, applying formal methods to a model working with image inputs would mean generating all adversarial images (within a certain noise range), feeding them to the model and verifying whether the output is as intended. Therefore, this class of countermeasures is still more theoretical than practical.

Empirical defenses, on the other hand, rely on experiments to ascertain the effectiveness of a defense mechanism. There are several defense strategies which can be employed. Adversarial training refers to retraining of the model with adversarial samples included in the training set after including their correct labels. It is expected that this will ensure that the model learn to ignore the noise and focus on the more evident features in the entire training set. A technique called Ensemble Adversarial Training (EAT) has been proposed that augments training data with perturbations transferred from other models, thereby making the model more robust (Tramèr et al. 2017). Cascade adversarial training, which transfers the knowledge of the end results of adversarial training on one model, to other models has been proposed to enhance the robustness of models (Na et al. 2017). A robust optimization based approach for identifying universally applicable, reliable training methods for neural networks has also been proposed (Madry et al. 2017).

Other commonly used technique to defend models against evasion attacks is input modification. In this case, an input sample, prior to being fed to the model, is passed through a sanitizing system to remove the adversarial noise, if any. Examples of such methods include denoising approaches like autoencoders and high level representational denoisers, JPEG compression, pixel deflection, and general basis function transformations (Moisejevs 2019). Lastly, there is an interesting NULL class approach (Hosseini et al. 2017), in which the classifier is trained to output a NULL class for inputs which it considers as adversarial.

6.4.3   Hardening Federated Learning Systems

Since the process of training, aggregation and model updating is spread over the client, server, and the network in a federated learning system, all the three segments need hardening against potential adversaries. Privacy protection at the client side may be ensured by adding perturbations (noise) to the updates (Ma et al. 2019). The more sensitive attributes in the update can be obscured by using differential privacy techniques (Dwork et al. 2006).

The server side can be made more robust by incorporating Secure Multi-Party Computation (SMC) which ensures that individual updates are rendered uninspectable at the server (Rosulek 2017). A secure aggregation protocol can be employed that uses cryptographic techniques so a coordinating server can only decrypt the average update if a certain number of users have participated, and no individual update can be inspected before averaging. A variety of other specialized approaches have also been employed to safeguard user privacy. These include, but are not limited to, de-identification schemes like anonymization, and cryptographic techniques like homomorphic encryption. In FL systems incorporating the latter, user updates are encrypted before uploading to the server using public-private keys (Papernot et al. 2016). Moreover, since the source of the updates is not required for the aggregation, the updates can be transferred without including metadata related to the origin of the information. Lastly, to safeguard against data poisoning attacks, anomaly detection schemes may be employed on the encrypted updates to identify any outliers, and the nodes which contributed those malicious samples may be removed from subsequent rounds of updates. Further, a weight may also be assigned to each user update based on its quality, and this process may help in identifying clients which are helpful in faster convergence or higher performance of the model. Conversely, clients with lower ranked updates may be identified as stragglers.

To make the actual communication of updates over a network more resilient to eavesdroppers, the client may also consider sending the updates over a mixed network like Tor, or via a trusted third party (Ma et al. 2019).

6.5   FUTURE DIRECTIONS

The previous sections presented an outline of the concept, applications and issues related to the emerging area of Edge AI. It was mentioned that although appealing, the incorporation of distributed intelligence in the edge devices is not without its share of limitations which need to be addressed before Edge AI can be said to be mature. This section presents an overview of the future research avenues in the field of Edge AI.

6.5.1   Open Issues in Federated Learning

As mentioned in the previous section, convergence in FL systems is still not theoretically proven. More research efforts are required towards improving learning performance, that is bettering learning accuracy with lesser communication between the edge devices and the centralized server. The present tradeoff between privacy preservation mechanisms and convergence speed needs further investigation to tilt the balance in favor of faster training with maximal user privacy. Recognition and prevention of data and model poisoning attacks is still an open problem, as is the security of the transmitted updates against eavesdroppers. Lastly, the process of aggregation may be made robust by incorporating mechanisms like anomaly detection to identify outliers (malicious updates). The use of reward functions for participating nodes is still in infancy, and needs more study. Incorporation of rewards into the FL system would provide incentives to devices contributing more to the learning process (either due to their having more data, or more computational capability). Lastly, the use of Blockchain has also been proposed to facilitate secure transmission of updates (Kim et al. 2018). However, blockchain based federated learning systems have yet to become mainstream.

6.5.2   Distributed Deep Reinforcement Learning

Reinforcement learning, being the closest ML algorithm to human learning in the sense that it learns from experience, is another technique which can be explored for improving the intelligence in edge devices. Such distributed Deep Reinforcement Learning (DRL) (also referred to as multi-agent DRL) is expected to bring revolutionary improvements in the way interconnected edge devices learn and infer. This assumes particular importance in Edge AI scenarios where most sensors participate in data generation without being able to obtain or assign class labels. Semi-Supervised DRL has already been proposed for such cases (Mohammadi et al. 2017), and Unsupervised DRL for incorporating learning in Edge AI systems with little to no supervision is another open area of research.

6.6   Conclusion

This chapter first presented a discussion on the security threats to conventional edge computing systems. Thereafter, techniques to incorporate intelligence into the edge devices were highlighted. This is pertinent since Edge

AI is ultimately expected to allow and encourage collaboration between various edge nodes towards a globally intelligent model without explicit human support. An overview of the various threats to the rapidly growing field of Edge AI was then presented. Security issues in various aspects of Edge AI were discussed and some effective countermeasures were highlighted. Lastly, avenues for future research in the area were outlined wherein it was discussed that emerging technologies like Blockchain and Deep Reinforcement Learning could be leveraged to improve existing Edge AI systems.

REFERENCES

Anley, Chris. 2002. *Advanced SQL Injection in SQL Server Applications.* Proceedings CGISecurity, 1–25.

Bagdasaryan, Eugene, Andreas Veit, Yiqing Hua, Deborah Estrin, and Vitaly Shmatikov. 2018. How to Backdoor Federated Learning. *arXiv preprint arXiv:1807.00459.*

Biggio, Battista, and Fabio Roli. 2018. Wild Patterns: Ten Years after the Rise of Adversarial Machine Learning. *Pattern Recognition* 84: 317–331.

Bogdanoski, Mitko, Tomislav Suminoski, and Aleksandar Risteski. 2013. Analysis of the SYN Flood DoS Attack. *International Journal of Computer Network and Information Security (IJCNIS)* 5 (8): 1–11.

Brendel, Wieland, Jonas Rauber, and Matthias Bethge. 2017. Decision-based Adversarial Attacks: Reliable Attacks Against Black-Box Machine Learning Models. *arXiv preprint arXiv:1712.04248.*

Carlini, Nicholas, and David Wagner. 2017. *Adversarial Examples are Not Easily Detected: Bypassing Ten Detection Methods.* Proceedings of the 10th ACM Workshop on Artificial Intelligence and Security, 3–14. ACM.

Carlini, Nicholas, Chang Liu, Úlfar Erlingsson, Jernej Kos, and Dawn Song. 2019. *The Secret Sharer: Evaluating and Testing Unintended Memorization in Neural Networks.* 28th {USENIX} Security Symposium ({USENIX} Security 19), 267–284.

Chen, Weiteng, and Zhiyun Qian. 2018. *Off-Path {TCP} Exploit: How Wireless Routers Can Jeopardize Your Secrets.* 27th {USENIX} Security Symposium ({USENIX} Security 18), 1581–1598.

Chen, Eric Y., Yutong Pei, Shuo Chen, Yuan Tian, Robert Kotcher, and Patrick Tague. 2014. *Oauth Demystified for Mobile Application Developers.* Proceedings of the 2014 ACM SIGSAC Conference on Computer and Communications Security, 892–903. ACM.

Chen, Pin-Yu, Huan Zhang, Yash Sharma, Jinfeng Yi, and Cho-Jui Hsieh. 2017a. *Zoo: Zeroth Order Optimization based Black-Box Attacks to Deep Neural Networks Without Training Substitute Models.* Proceedings of the 10th ACM Workshop on Artificial Intelligence and Security, 15–26. ACM.

Chen, Xinyun, Chang Liu, Bo Li, Kimberly Lu, and Dawn Song. 2017b. Targeted Backdoor Attacks on Deep Learning Systems Using Data Poisoning. *arXiv preprint arXiv:1712.05526.*

Chen, Pin-Yu, Yash Sharma, Huan Zhang, Jinfeng Yi, and Cho-Jui Hsieh. 2018a. *EAD: Elastic-Net Attacks to Deep Neural Networks Via Adversarial Examples.* Thirty-Second AAAI Conference on Artificial Intelligence.

Chen, Yimin, Tao Li, Rui Zhang, Yanchao Zhang, and Terri Hedgpeth. 2018b. *Eyetell: Video-Assisted Touchscreen Keystroke Inference from Eye Movements.* 2018 IEEE Symposium on Security and Privacy (SP), 144–160. IEEE.

Cheng, Yu, Duo Wang, Pan Zhou, and Tao Zhang. 2017. A Survey of Model Compression and Acceleration for Deep Neural Networks. *arXiv preprint arXiv:1710.09282.*

Chollet, François. 2017. *Xception: Deep Learning with Depthwise Separable Convolutions.* Proceedings of the IEEE Conference on Computer Vision and Pattern Recognition, 1251–1258.

Cisco. 2016. Cisco Fog Director Cross-Site Scripting Vulnerability. Accessed 22 October 2019. https://tools.cisco.com/security/center/content/CiscoSecurityAdvisory/cisco-sa-20160201-fd.

Costin, Andrei. 2018. *IoT/Embedded vs. Security: Learn from the Past, Apply to the Present, Prepare for the Future.* Proceedings of Conference of Open Innovations Association FRUCT. FRUCT Oy.

Cretu, Gabriela F., Angelos Stavrou, Michael E. Locasto, Salvatore J. Stolfo, and Angelos D. Keromytis. 2008. *Casting Out Demons: Sanitizing Training Data for Anomaly Sensors.* 2008 IEEE Symposium on Security and Privacy (SP 2008), 81–95. IEEE.

Dalvi, Nilesh, Pedro Domingos, Sumit Sanghai, and Deepak Verma. 2004. *Adversarial Classification.* Proceedings of the Tenth ACM SIGKDD International Conference on Knowledge Discovery and Data Mining, 99–108. ACM.

Damon, Evan, Julian Dale, Evaristo Laron, Jens Mache, Nathan Land, and Richard Weiss. 2012. *Hands-on Denial of Service Lab Exercises Using Slowloris and Rudy.* Proceedings of the 2012 Information Security Curriculum Development Conference, 21–29. ACM.

Das, Dipankar, Sasikanth Avancha, Dheevatsa Mudigere, Karthikeyan Vaidynathan, Srinivas Sridharan, Dhiraj Kalamkar, Bharat Kaul, and Pradeep Dubey. 2016. Distributed Deep Learning Using Synchronous Stochastic Gradient Descent. *arXiv preprint arXiv:1602.06709.*

Dean, Jeffrey, Greg Corrado, Rajat Monga, Kai Chen, Matthieu Devin, Mark Mao, Marc'aurelio Ranzato et al. 2012. Large Scale Distributed Deep Networks. *Advances in Neural Information Processing Systems,* 1223–1231.

Dhanapal, A., and P. Nithyanandam. 2017. *An Effective Mechanism to Regenerate HTTP Flooding DDoS Attack Using Real Time Data Set.* 2017 International Conference on Intelligent Computing, Instrumentation and Control Technologies (ICICICT), 570–575. IEEE.

Dwork, Cynthia, Frank McSherry, Kobbi Nissim, and Adam Smith. 2006. *Calibrating Noise to Sensitivity in Private Data Analysis.* Theory of Cryptography Conference, 265–284. Berlin, Heidelberg: Springer.

Enev, Miro, Sidhant Gupta, Tadayoshi Kohno, and Shwetak N. Patel. 2011. *Televisions, Video Privacy, and Powerline Electromagnetic Interference.* Proceedings of the 18th ACM Conference on Computer and Communications Security, 537–550. ACM.

Fredrikson, Matt, Somesh Jha, and Thomas Ristenpart. 2015. *Model Inversion Attacks that Exploit Confidence Information and Basic Countermeasures.* Proceedings of the 22nd ACM SIGSAC Conference on Computer and Communications Security, 1322–1333. ACM.

Gao, Yansong, Chang Xu, Derui Wang, Shiping Chen, Damith C. Ranasinghe, and Surya Nepal. 2019. STRIP: A Defence Against Trojan Attacks on Deep Neural Networks. *arXiv preprint arXiv:1902.06531.*

Greenberg, Andy. 2017. The Reaper IoT Botnet has Already Infected a Million Networks. Accessed 13 January 2018. https://www.wired.com/story/reaper-iot-botnet-infected-million-networks/.

Hosseini, Hossein, Yize Chen, Sreeram Kannan, Baosen Zhang, and Radha Poovendran. 2017. Blocking Transferability of Adversarial Examples in Black-Box Learning Systems. *arXiv preprint arXiv:1703.04318.*

Howard, Andrew G., Menglong Zhu, Bo Chen, Dmitry Kalenichenko, Weijun Wang, Tobias Weyand, Marco Andreetto, and Hartwig Adam. 2017. Mobilenets: Efficient Convolutional Neural Networks for Mobile Vision Applications. *arXiv preprint arXiv:1704.04861.*

Iandola, Forrest N., Song Han, Matthew W. Moskewicz, Khalid Ashraf, William J. Dally, and Kurt Keutzer. 2016. SqueezeNet: AlexNet-Level Accuracy with 50x Fewer Parameters and <0.5 MB Model Size. *arXiv preprint arXiv:1602.07360.*

Ilyas, Andrew, Logan Engstrom, Anish Athalye, and Jessy Lin. 2018. Black-Box Adversarial Attacks with Limited Queries and Information. *arXiv preprint arXiv:1804.08598.*

Iorga, Michaela, Larry Feldman, Robert Barton, Michael J. Martin, Nedim S. Goren, and Charif Mahmoudi. 2018. *Fog Computing Conceptual Model.* No. Special Publication (NIST SP)-500-325.

Jagielski, Matthew, Alina Oprea, Battista Biggio, Chang Liu, Cristina Nita-Rotaru, and Bo Li. 2018. *Manipulating Machine Learning: Poisoning Attacks and Countermeasures for Regression Learning.* 2018 IEEE Symposium on Security and Privacy (SP), 19–35. IEEE.

Kim, Hyesung, Jihong Park, Mehdi Bennis, and Seong-Lyun Kim. 2018. On-Device Federated Learning Via Blockchain and Its Latency Analysis. *arXiv preprint arXiv:1808.03949.*

Konečný, Jakub, H. Brendan McMahan, Felix X. Yu, Peter Richtárik, Ananda Theertha Suresh, and Dave Bacon. 2016. Federated Learning: Strategies for Improving Communication Efficiency. *arXiv preprint arXiv:1610.05492.*

Krizhevsky, Alex, Ilya Sutskever, and Geoffrey E. Hinton. 2012. Imagenet Classification with Deep Convolutional Neural Networks. *Advances in Neural Information Processing Systems* 25: 1097–1105.

Kurakin, Alexey, Ian Goodfellow, and Samy Bengio. 2016. Adversarial Examples in the Physical World. *arXiv preprint arXiv:1607.02533*.

Ma, Chuan, Jun Li, Ming Ding, Howard Hao Yang, Feng Shu, Tony Q.S. Quek, and H. Vincent Poor. 2019. On Safeguarding Privacy and Security in the Framework of Federated Learning. *arXiv preprint arXiv:1909.06512*.

Madry, Aleksander, Aleksandar Makelov, Ludwig Schmidt, Dimitris Tsipras, and Adrian Vladu. 2017. Towards Deep Learning Models Resistant to Adversarial Attacks. *arXiv preprint arXiv:1706.06083*.

McIntosh, Michael, and Paula Austel. 2005. *XML Signature Element Wrapping Attacks and Countermeasures*. Proceedings of the 2005 Workshop on Secure Web Services, 20–27. ACM.

Mei, Shike, and Xiaojin Zhu. 2015. *Using Machine Teaching to Identify Optimal Training-Set Attacks on Machine Learners*. Twenty-Ninth AAAI Conference on Artificial Intelligence.

Mohammadi, Mehdi, Ala Al-Fuqaha, Mohsen Guizani, and Jun-Seok Oh. 2017. Semisupervised Deep Reinforcement Learning in Support of IoT and Smart City Services. *IEEE Internet of Things Journal* 5 (2): 624–635.

Moisejevs, Ilja. 2019. Evasion Attacks on Machine Learning (or "Adversarial Examples"). Accessed 22 October 2019. https://towardsdatascience.com/evasion-attacks-on-machine-learning-or-adversarial-examples-12f2283e06a1.

Na, Taesik, Jong Hwan Ko, and Saibal Mukhopadhyay. 2017. Cascade Adversarial Machine Learning Regularized with a Unified Embedding. *arXiv preprint arXiv:1708.02582*.

Nakhila, Omar, Afraa Attiah, Yier Jin, and Cliff Zou. 2015. *Parallel Active Dictionary Attack on wpa2-psk wi-fi Networks*. MILCOM 2015-2015 IEEE Military Communications Conference, 665–670. IEEE.

National Institute of Standards and Technology. 2010. National Vulnerability Database CVE-2010-3972 Detail. Accessed 22 October 2019. https://nvd.nist.gov/vuln/detail/CVE-2010-3972.

Nelson, Blaine, Marco Barreno, Fuching Jack Chi, Anthony D. Joseph, Benjamin I.P. Rubinstein, Udam Saini, Sutton Charles, J.D. Tygar, and Kai Xia. 2009. Misleading Learners: Co-opting Your Spam Filter. In *Machine Learning in Cyber Trust*, 17–51. Boston, MA: Springer.

Örs, Sıddıka Berna, Elisabeth Oswald, and Bart Preneel. 2003. *Power-Analysis Attacks on an FPGA—First Experimental Results*. International Workshop on Cryptographic Hardware and Embedded Systems, 35–50. Berlin, Heidelberg: Springer.

Papernot, Nicolas, Martín Abadi, Ulfar Erlingsson, Ian Goodfellow, and Kunal Talwar. 2016. Semi-supervised Knowledge Transfer for Deep Learning from Private Training Data. *arXiv preprint arXiv:1610.05755*.

Paudice, Andrea, Luis Muñoz-González, Andras Gyorgy, and Emil C. Lupu. 2018. Detection of Adversarial Training Examples in Poisoning Attacks Through Anomaly Detection. *arXiv preprint arXiv:1802.03041.*

Roman, Rodrigo, Javier Lopez, and Masahiro Mambo. 2018. Mobile Edge Computing, Fog et al.: A Survey and Analysis of Security Threats and Challenges. *Future Generation Computer Systems* 78: 680–698.

Rosulek, Mike. 2017. *Improvements for Gate-Hiding Garbled Circuits.* International Conference on Cryptology in India, 325–345. Cham: Springer.

Sandler, Mark, Andrew Howard, Menglong Zhu, Andrey Zhmoginov, and Liang-Chieh Chen. 2018. *Mobilenetv2: Inverted Residuals and Linear Bottlenecks.* Proceedings of the IEEE Conference on Computer Vision and Pattern Recognition, 4510–4520.

Shi, Weisong, Jie Cao, Quan Zhang, Youhuizi Li, and Lanyu Xu. 2016. Edge Computing: Vision and Challenges. *IEEE Internet of Things Journal* 3 (5): 637–646.

Shokri, Reza, Marco Stronati, Congzheng Song, and Vitaly Shmatikov. 2017. *Membership Inference Attacks Against Machine Learning Models.* 2017 IEEE Symposium on Security and Privacy (SP), 3–18. IEEE.

Sonar, Krushang, and Hardik Upadhyay. 2014. A Survey: DDOS Attack on Internet of Things. *International Journal of Engineering Research and Development* 10 (11): 58–63.

Steinhardt, Jacob, Pang Wei W. Koh, and Percy S. Liang. 2017. Certified Defenses for Data Poisoning Attacks. In *Neural Information Processing Systems Foundation, Inc.*, Long Beach, 4–9, December 2017, 3517–3529, USA.

Suciu, Octavian, Radu Marginean, Yigitcan Kaya, Hal Daume III, and Tudor Dumitras. 2018. *When Does Machine Learning {FAIL}? Generalized Transferability for Evasion and Poisoning Attacks.* 27th {USENIX} Security Symposium ({USENIX} Security 18), 1299–1316.

Sun, San-Tsai, and Konstantin Beznosov. 2012. *The Devil is in the (Implementation) Details: An Empirical Analysis of OAuth SSO Systems.* Proceedings of the 2012 ACM Conference on Computer and Communications Security, 378–390. ACM.

Szegedy, Christian, Wojciech Zaremba, Ilya Sutskever, Joan Bruna, Dumitru Erhan, Ian Goodfellow, and Rob Fergus. 2013. Intriguing Properties of Neural Networks. *arXiv preprint arXiv:1312.6199.*

Szegedy, Christian, Wei Liu, Yangqing Jia, Pierre Sermanet, Scott Reed, Dragomir Anguelov, Dumitru Erhan, Vincent Vanhoucke, and Andrew Rabinovich. 2015. *Going Deeper with Convolutions.* Proceedings of the IEEE Conference on Computer Vision and Pattern Recognition, 1–9.

Tramèr, Florian, Alexey Kurakin, Nicolas Papernot, Ian Goodfellow, Dan Boneh, and Patrick McDaniel. 2017. Ensemble Adversarial Training: Attacks and Defenses. *arXiv preprint arXiv:1705.07204.*

Truex, Stacey, Ling Liu, Mehmet Emre Gursoy, Lei Yu, and Wenqi Wei. 2019. Demystifying Membership Inference Attacks in Machine Learning as a Service. *IEEE Transactions on Services Computing.* https://ieeexplore.ieee.org/abstract/document/8634878

Udhayan, J., and R. Anitha. 2009. *Demystifying and Rate Limiting ICMP Hosted DoS/DDoS Flooding Attacks with Attack Productivity Analysis.* 2009 IEEE International Advance Computing Conference, 558–564.

Uesato, Jonathan, Brendan O'Donoghue, Aaron van den Oord, and Pushmeet Kohli. 2018. Adversarial Risk and the Dangers of Evaluating Against Weak Attacks. *arXiv preprint arXiv:1802.05666.*

Vanhoef, Mathy, and Frank Piessens. 2018. *Release the Kraken: New KRACKs in the 802.11 Standard.* Proceedings of the 2018 ACM SIGSAC Conference on Computer and Communications Security, 299–314. ACM.

Xiao, Yinhao, Yizhen Jia, Chunchi Liu, Xiuzhen Cheng, Jiguo Yu, and Weifeng Lv. 2019. Edge Computing Security: State of the Art and Challenges. *Proceedings of the IEEE* 107 (8): 1608–1631.

Xiaoming, Li, Valon Sejdini, and Hasan Chowdhury. 2010. Denial of Service (dos) Attack with UDP Flood. *School of Computer Science, University of Windsor, Canada.*

Zhang, Shanshan, Ce Zhang, Zhao You, Rong Zheng, and Bo Xu. 2013. *Asynchronous Stochastic Gradient Descent for DNN Training.* 2013 IEEE International Conference on Acoustics, Speech and Signal Processing, 6660–6663. IEEE.

Zhou, Xiaoyong, Soteris Demetriou, Dongjing He, Muhammad Naveed, Xiaorui Pan, XiaoFeng Wang, Carl A. Gunter, and Klara Nahrstedt. 2013. *Identity, Location, Disease and More: Inferring Your Secrets from Android Public Resources.* Proceedings of the 2013 ACM SIGSAC Conference on Computer & Communications Security, 1017–1028. ACM.

CHAPTER 7

Examining Privacy Disclosure and Trust in the Consumer Internet of Things: An Integrated Research Framework

Grace Fox and Theo Lynn

Abstract The Internet of Things (IoT) and the various applications it encompasses offer great potential for personalisation and convenience in all aspects of individuals' lives from healthcare to transport and smart homes. However, IoT devices collect and share large volumes of personal data leading to concerns for the security and privacy of the data. While computer science research has explored technical solutions to security issues, it is important to explore privacy from the perspective of consumers. To foster a sense of privacy and trust among consumers, IoT service providers must communicate with consumers regarding their data practices in a transparent manner. To do this, we propose that IoT service providers refine adopt transparent privacy disclosure approaches. We present a framework for testing the effectiveness of privacy disclosures in building consumers' perceptions of privacy and trust and empowering consumers to adopt IoT devices whilst retaining some level of privacy. We illustrate this framework with reference to a privacy label approach.

G. Fox (✉) • T. Lynn
Irish Institute of Digital Business, DCU Business School, Dublin, Ireland
e-mail: grace.fox@dcu.ie; theo.lynn@dcu.ie

© The Author(s) 2020   123
T. Lynn et al. (eds.), *The Cloud-to-Thing Continuum*, Palgrave
Studies in Digital Business & Enabling Technologies,
https://doi.org/10.1007/978-3-030-41110-7_7

Keywords Privacy • Trust • Privacy label • Trust label • Social contract theory • Information–Motivation–Skills Model • Research framework

7.1   Introduction

We now live in a world with more connected devices than people. In the near future, the Internet of Things (IoT) landscape will comprise of billions of connected devices and things with the ability to exchange data at any given time. IoT can be defined as

> A world where physical objects are seamlessly integrated into the information network, and where the physical objects can become active participants in business processes. Services are available to interact with these "smart objects" over the Internet, query their state and any information associated with them, taking into account security and privacy issues. (Haller et al. 2008, p. 15)

The potential value of IoT is enormous ranging from US$3.9 trillion to US$19 trillion in the coming years (Cisco 2013a, b; McKinsey Global Institute 2015). Notwithstanding this massive economic opportunity, IoT and the big data it generates further complicate the issues around privacy and security (Lowry et al. 2017). The connection of devices enabled by IoT can heighten privacy and security challenges, not least excessive monitoring and data mining techniques that may enable data to be made available for purposes for which it was not previously intended (Abomhara and Køien 2014). The risks associated with these challenges is exacerbated by the long service chains inherent in the Internet of Things involving a multitude of actors including not only IoT software vendors and device manufacturers but network operators, cloud service providers, and the software and hardware vendors and services to support the infrastructure underlying the IoT. While consumers may accept a degree of consumer surveillance from the Internet or IoT, they may be equally ignorant about the degree to which their data is being distributed to fulfil their service requirements. There is an onus on enterprises providing IoT products and services, and consuming IoT data, to both take privacy preserving actions and to communicate with consumers on the use of their data in the Internet of Things.

While existing research has identified some solutions to security challenges in IoT, user privacy and issues around privacy in data collection, management, and dissemination must be addressed (Abomhara and Køien 2014). Indeed, privacy and trust are categorised as two of the core security challenges facing the future of IoT (Sicari et al. 2015). Chapter 6 discusses some of the technical challenges at play in relation to the security of data. In this chapter, we focus on exploring the issues of privacy and trust related to IoT from the perspective of consumers. The remainder of the chapter is structured as follows; the next section explores perspectives and theories on privacy and the Internet of Things. It is followed by a brief discussion on the nature of trust and trust in technology. Next, we discuss approaches for influencing perceptions of privacy and trust. Following on from this literature, we propose an IoT privacy trust label as a potential means to influence perceptions and trust in relation to IoT. Based on theories, constructs, and concepts discussed in earlier sections, we present a framework for testing the effectiveness of privacy disclosures in building consumers' perceptions of privacy and trust and empowering consumers to adopt IoT devices whilst retaining some level of privacy. We illustrate this framework with reference to a privacy label approach.

7.2   Privacy and the Internet of Things

Users' privacy remains an important issue in IoT environments with concerns raised around the leakage of location information and inferences from IoT device usage such as TVs (Alrawais et al. 2017). It would seem while parents were once worried about the amount of time kids spent watching television, we now need to worry about the amount of time our television spends watching us.

In the context of IoT, there are several dimensions of privacy that must be considered and protected. These include identity data or personally identifiable information (PII), location data which can reveal many forms of PII, footprint privacy, and data contained in queries (Daubert et al. 2015). Solutions have been identified for many of these dimensions, such as anonymisation (Daubert et al. 2015), but again these solutions are technical in nature and do not emphasise the user perceptions. When focusing on user privacy, there is a tendency to focus on the application layer as this is the layer closest to the consumer and the point at which privacy perceptions can be addressed. It is also important to explore consumers' perceptions of privacy and trust as research has shown concern for privacy and absence of trust can both reduce willingness to disclose information and adopt new technologies (Li 2012).

Privacy as a phenomenon has been studied for centuries across a range of academic disciplines and perspectives such as law, sociology, marketing, and information systems (IS). This chapter focuses on privacy from the IS perspective. Privacy is defined as an individual's desire for greater control over the collection and dissemination of their personal information (Bélanger and Crossler 2011). This definition remains relevant in the context of IoT, with privacy described in this chapter as consumers' desire to be afforded a greater degree of control over the collection and use of their personal data by IoT devices and sensors. The IS literature on privacy has grown over the past three decades but privacy remains relevant today with polls continuing to find that individuals place value on their privacy. For example, Pike et al. (2017) found that 84% of consumers in the US expressed data privacy concerns, 70% of whom felt these concerns had recently increased. This may be attributable in part to growing awareness of incidences of data breaches, but it is likely to be in part influenced by the ever increasing volume of data collection facilitated by the growing proliferation of technology such as IoT devices.

Extant privacy research in the IS domain leverages a number of theoretical lenses to understand the role of privacy across different contexts and information technologies. In his review of the literature Li (2012) categorises privacy theories into five areas of theories focused on; (1) drivers of privacy concern, (2) behavioural consequences, (3) trade-offs, (4) institutional drivers and (5) individual factors. While the privacy literature in the IoT domain is in a nascent stage, the existing literature focuses on theories related to behavioural consequences, trade-offs, and individual factors to a lesser degree. In terms of behavioural consequences, many of the existing IoT studies leverage technology adoption models such as the theory of reasoned action (TRA) (Marakhimov and Joo 2017). These studies build understanding of the factors driving individuals' initial adoption decision making process, but do not enhance understanding of individuals' post-use behaviours and barriers to the use of IoT (Marakhimov and Joo 2017).

One dominant stream of the broader privacy literature focuses on the trade-offs consumers make between the benefits and risks associated with new technology use and as a result information disclosure. The relevance of trade-offs are apparent in the IoT context. As the number of devices a user connects with increases, the convenience and perceived benefits this usage facilitates increase (Hsu and Lin 2016) enabling users to query anything from health data to weather or utility usage. The data generated from the various IoT devices and connected databases does offer benefits but also introduces undeniable risks to consumers' privacy (Bélanger and Xu 2015). The most common theory to explore these trade-offs is the privacy calculus theory, which posits that individuals will disclose their personal information or interact with a technology for as long as the perceived benefits outweigh the perceived risks or consequences (Culnan 1993). The theory assumes that individuals conduct a cognitive cost-benefit analysis, considering the benefits of disclosure and the potential negative outcomes or repercussions the individual might experience as a result of using the technology (Culnan and Armstrong 1999). PCT has been recently leveraged in the IoT context. In their study of 508 Taiwanese citizens, Hsu and Lin (2016) found concern for information privacy had a negative influence on intentions to continue use of IoT, whereas perceived benefits had a positive influence on intentions. In a study of US consumers, Kim et al. (2019) explored perceptions of trust and benefits and perceived risk on three IoT services namely healthcare, smart home, and smart transport. In terms of healthcare, privacy risk had a significant negative influence on willingness to disclosure personal data, with trust and perceived benefits positively influencing willingness. In terms of both smart transport and smart homes, trust and perceived benefits had a significant, positive effect but perceived risk was insignificant. Perceived benefits was the biggest predictor of willingness to provide information in the case of healthcare and smart transport, whereas trust was the biggest predictor in the case of smart homes. These studies provide empirical support for the use of PCT in the IoT context, illustrating that both positive perceptions (i.e. trust and benefits) influence adoption and information disclosure, and negative perceptions (i.e. risk and privacy concern) can have a negative influence.

Notwithstanding the foregoing, due to biases and cognitive limitations, consumer's perception of the benefits often outweighs perceived risks or concerns. This view has also been presented in the IoT context with Kim et al. (2019) arguing that consumers seek benefits in spite of their privacy concerns and often underestimate the risks of IoT usage to their data privacy. This contradiction is termed the 'privacy paradox'. However, research explaining the privacy paradox is still emerging. Furthermore, it is important to consider potential knowledge gaps (Crossler and Bélanger 2017). Individuals may assume their data remains private and is not shared with other parties (Kim et al. 2019), and thus their behaviours may only seem to contradict their desire for privacy. Furthermore, we do not yet fully understand how behaviours contradict privacy concerns (Keith et al. 2015).

In terms of individual theories, protection motivation theory (PMT) is frequently leveraged in the privacy literature to explore the influence of individuals' threat and coping appraisals on their behaviours (Li 2012). In their study of 206 health wearable users in the United States, Marakhimov and Joo (2017) leverage PMT. They found that consumers' threat appraisal was significantly influenced by their general privacy concerns and their health information privacy concerns, with threat appraisals significantly influencing problem and emotion focused coping and extended use intentions as a result.

In the IoT context, no study has yet explored privacy using an institutional based-theory. However, in their early stage work, Saffarizadeh et al. (2017) leverage social reciprocity theory to propose a model which explains consumers' willingness to disclose personal data to conversational assistants. They include privacy concerns as a negative determinant on disclosure. As perceived trustworthiness leads to consumers being more likely to disclose information (McKnight et al. 2011), to foster this trust, Saffarizadeh et al. (2017) argue that in line with social reciprocity theory, disclosures from conversational assistants may encourage users to trust them. These studies provide important insights into the perceptions driving behaviour in the IoT context, but it is important to explore approaches to influence these perceptions and engender perceptions of trust and privacy as a result.

7.3   Trust, Privacy, and the Internet of Things

A consumer's willingness to trust is based on their beliefs of the trustworthiness of the organisation (van der Werff et al. 2019). These beliefs together encapsulate the assumption that the organisation will not engage in opportunistic behaviour with the individual's data (Dinev and Hart 2006) and generally relate to beliefs regarding the organisation's benevolence, integrity, and competence (van der Werff et al. 2019). Benevolence relates to the belief the organisation has the individual's best interests in mind, integrity refers to the belief in the morals and principles of the organisation, and competence refers to the belief the organisation has the knowledge and skills to fulfil a service (Belanger et al. 2002).

Trust and privacy are often studied in tandem in many contexts including IoT, with privacy concerns negatively impacting disclosure or technology adoption and trust having the opposite influence. Generally speaking, trust in a privacy context relates to an individual's willingness to be vulnerable when transacting or sharing personal information with an organisation (McKnight et al. 2011). In the IoT context, trust can be described as consumers' willingness to be vulnerable when interacting and sharing personal data with an IoT device, the associated application, and the organisation(s) providing these. In the IoT context, there are also dimensions of trust to consider namely device trust, processing trust, connection trust to ensure data is exchanged appropriately and trust in the overall system (Daubert et al. 2015). The opaqueness of the IoT service chain makes this logistically near-impossible. While there are technical solutions in place or proposed to achieve these dimensions of trust such as trusted computing, confidentiality, certifications, and more recently, blockchain (Daubert et al. 2015; Chanson et al. 2019), there is a need to account for consumers' perceptions of trustworthiness.

7.4   Approaches for Influencing Perceptions of Privacy and Trust

As evidenced in the IoT and broader privacy literature, concern for privacy negatively impacts disclosure and willingness to use new technologies, whereas trust can positively impact adoption and disclosure behaviours (Kim et al. 2019). However, the nature of the Internet and interactions between consumers and technology or devices complicates mechanisms for building trust (van der Werff et al. 2019). It is thus important to explore mechanisms to build a sense of privacy, that is perceived control over how one's personal information is collected and used, and foster a sense of trust, that is consumers' willingness to accept vulnerability when interacting with IoT devices.

In terms of overcoming privacy concerns, prevailing suggestions in the privacy literature include increasing consumers' perceptions of control (Tucker 2014), building trust (Dinev and Hart 2006) and reducing perceptions of risk (Xu et al. 2011). In order to influence consumers' perceptions, organisations must transparently communicate with users with regards to the controls they have over their personal data, what data is collected, and how data is used. While the efficacy of organisations' communication methods in the IoT context is yet to be tested, the need for communication prevails. For instance, in a study of smartwatch users, Williams et al. (2019) found that users who had not been primed on the risks to their personal data on smartwatches, did not perceive any risks as they hadn't learned the value of this data. Researchers have proposed that IoT providers offer users an awareness of the privacy risks, provide users with control over the collection and usage of their data by smart devices (Ziegeldorf et al. 2014; Davies et al. 2016), and control over subsequent usage by additional third-party entities and devices (Hsu and Lin 2016). This again highlights the importance of education efforts for users of IoT devices.

In terms of trust, there are no means to assess trustworthiness of IoT devices (Alrawais et al. 2017). Trust is typically developed over time as opposed to being formed based on a one-time interaction (Gefen et al. 2008). This makes trust building between consumers and online organisations or IoT devices complex. To build trust in online organisations, several approaches have been explored. Firstly the characteristics of a website such as website design, security seals or privacy policies have been examined in the literature (van der Werff et al. 2018). However, the findings on the effectiveness of these approaches have been mixed. Moreover, given that the interaction with IoT devices does not involve regular interaction with websites, many of these methods are impractical or insufficient. It is also important for the user to trust the device, as highlighted in the study by Saffarizadeh et al. (2017), and the organisation itself (IoT service provider).

The dominant method for communicating how organisations collect and use consumers' data are privacy policies. It is argued that privacy policies could reduce perceived risks, increase perceptions of control and trust (Xu et al. 2011; Pan and Zinkhan 2006) and thereby overcome any privacy obstacles. However, privacy policies tend to be quite lengthy and difficult to read (Kelley et al. 2010). Thus, when customers read privacy policies, they fail to understand the contents (Park et al. 2012) and as a result these disclosures may have the opposite to the intended impact and exacerbate concerns around control and risk. There is a need to both adjust the content of policies and develop methods which better inform consumers of how their information is used (Park et al. 2012). To combat these issues, researchers developed the privacy label based on the nutrition

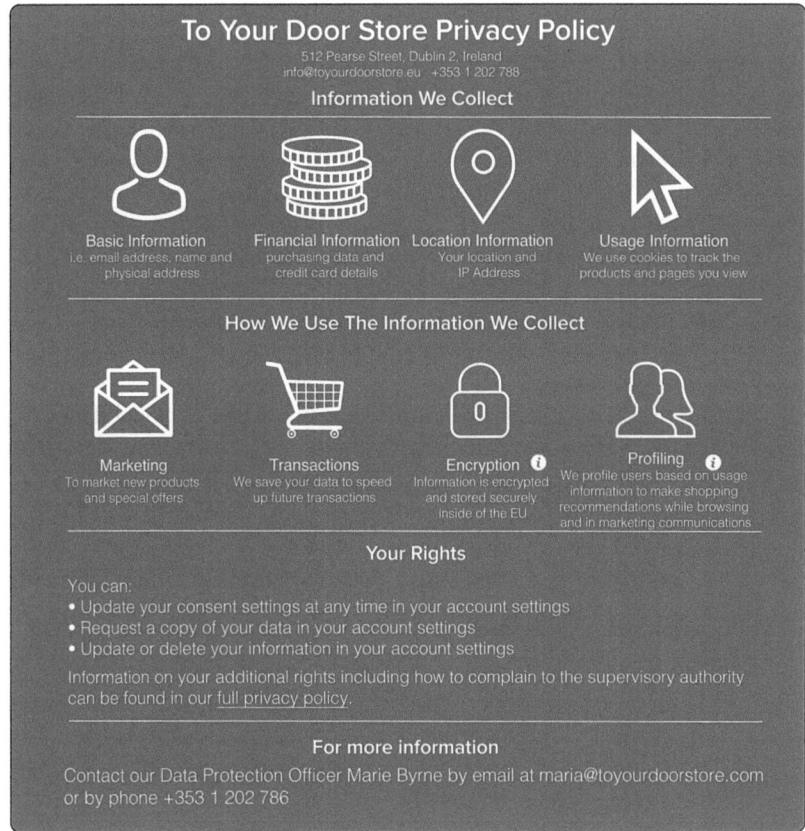

Fig. 7.1 Example GDPR label (Fox et al. 2018)

label approaches and found that privacy labels could improve understanding of privacy practices (Kelley et al. 2009, 2010) and build perceptions of trust (van der Werff et al. 2019). This approach has recently been adapted to develop GDPR-based (General Data Protection Regulation) privacy labels (see Fig. 7.1) (Fox et al. 2018).

7.5   Privacy Trust Labels: Design Principles

We argue that IoT service providers should draw from this recent research on privacy and trust labels to develop an IoT based privacy label. The label should seek to build consumers' understanding of how their data is used and collected to comply with privacy regulation and build positive privacy perceptions, as well as information on the organisation to build perceptions of trustworthiness. For example, in Europe, to comply with the GDPR, the labels must include the following information (ICO 2017):

1. The identity and contact details of the data controller
2. The processing purposes for the personal data and the legal basis for the processing
3. The recipients or categories of recipients of the personal data
4. The details of the safeguards in place if transferring data to a third country
5. Data retention period
6. The data subject's rights to request: access to their data, rectification, restriction of processing, erasure of data, and data portability
7. If data processing is based on consent, the right to withdraw consent at any time
8. The right to complain to the supervisory authority
9. whether the disclosure of personal data is a statutory or contractual requirement and the consequences of non-disclosure
10. The use of automated decision-making such as profiling, the logic and impact of such processing
11. The contact details of the data protection officer
12. Information on further processing.

All information on the label should be framed in a manner, which demonstrates the benevolence, integrity, and competence of the IoT service provider with regards to protecting consumers' personal data. Traditionally, privacy labels are presented to users upon signing up to an online website or service. As IoT devices cross physical and informational boundaries, the physical security and wellbeing of citizens and their homes is intertwined in the security and privacy of the IoT devices and the network (Lowry et al. 2017). We thus, recommend the inclusion of physical privacy labels on the box of IoT devices, along with a digital label on the application presented to users at sign-up and an up to date label accessible within the application's privacy features and on the service provider's website.

7.6   Towards a Framework for Examining the Impact of Privacy Disclosures on Privacy Perceptions and Behaviours

In this section, we present a general framework for building consumers' perceptions of trust and privacy in the IoT context in Fig. 7.2 below that can be used for examine privacy and trust perceptions and behaviours in the Internet of Things. We illustrate the use of this framework in the context of the Privacy Trust Label described in Sect. 7.5 above.

With IoT technologies advancing at a faster pace than privacy regulation and practices (Lowry et al. 2017), it is important for IoT service providers to be proactive in addressing consumers' privacy concerns. Consumer perceptions of privacy are situational in nature in that they are influenced by past experience and the context in question (Li 2011). For example, individuals have perceptions of how much privacy they have in the e-commerce context, which may be influenced by past experience of a positive nature, such as convenient online shopping, and experience of a negative nature, such as a privacy invasion. Furthermore, individuals' have perceptions regarding well-known brands. These perceptions may relate to how the brand protects consumer privacy and how trustworthy the brand is with regards to protecting and fairly using personal data. For example, if a consumer perceives that Apple smartphones offer a

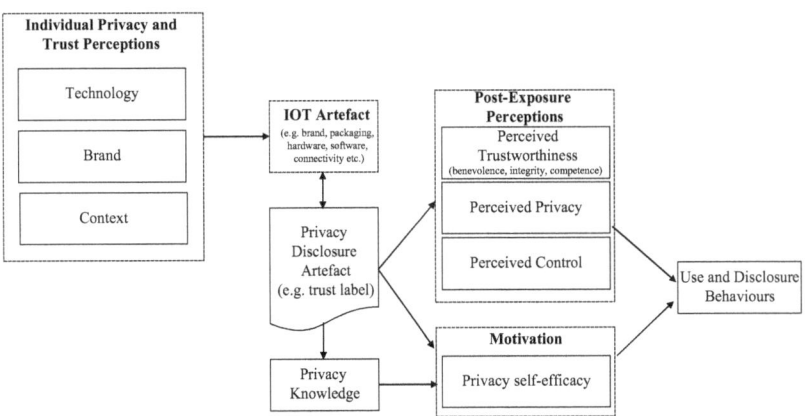

Fig. 7.2 Integrated framework to examining privacy and trust perceptions and behaviours

satisfactory level of privacy and the brand is trustworthy in terms of competence to protect data, integrity and benevolence with how that data is used, the consumer may hold positive perceptions about the trustworthiness and privacy offered by Apple products in other contexts such as the Apple watch or Apple TV. We present a framework that recognises that consumers have pre-existing perceptions and preferences regarding privacy and trust in technologies, brand and contexts (e.g. health, finance, social media, etc.). These may be general perceptions and preferences or specific to IoT. As such, these perceptions and preferences influence and are influenced by the brand, packaging, and the device hardware, software, and connectivity.

We draw from the integrative privacy framework developed by Li (2012) and the recently adapted Information–Motivation–Behavioural Skills Model by Crossler and Bélanger (2019). On the left hand of the model the IoT privacy label is presented. The label will seek to build consumers' privacy knowledge regarding how their personal data is collected and used by IoT devices. This label will in turn influence consumers' perceptions regarding the IoT device and service provider. In line with social contract theory (SCT) theory, we argue that the label will foster perceptions of control, trustworthiness, and privacy. SCT proposes that when organisations engage in transactions with customers which involve the disclosure of personal data, they enter into a social contract (Donaldson and Dunfee 1994). This contract implies that the organisation will only use the personal data in ways which align with social norms and that individuals have some level of control (Bélanger and Crossler 2011). We argue that the privacy label will form the basis of a social contract informing consumers of how their personal data is collected, stored, and disseminated in this specific context of the IoT device. Previous research has shown that privacy disclosures can enhance perceived control (Xu et al. 2011). We therefore argue that if consumers believe they retain some level of control over their personal data, they are more likely be willing to use IoT devices and disclose personal data. Similarly, privacy disclosures can potentially lead individuals to form positive perceptions related to privacy and heighten individuals' beliefs in the trustworthiness of the organisation (Culnan and Armstrong 1999). We propose a similar effect in the context of IoT devices.

Following on from perceptions and knowledge, Crossler and Bélanger (2019) discuss the privacy knowledge–belief gap and highlight the importance of contextualised privacy self-efficacy, that is individuals' perceptions that they have the knowledge and skills needed to protect the privacy of their data as required. We argue that the privacy label will provide context-specific insights into *how* users can retain control over their data collected by IoT devices. This self-efficacy will in turn influence consumers' intentions to engage in privacy-protective behaviours such as adapting privacy settings (Crossler and Bélanger 2017, 2019). On the right hand of the model is users' usage and disclosure behaviours. We argue that the privacy label will build consumers' privacy self-efficacy and provide them with the motivation to exercise control over their privacy by modifying the privacy settings on IoT devices. We propose that consumers with high self-efficacy will adopt and continue to use IoT devices due to the high perceptions of control, privacy, and trust fostered by the label and reconfirmed through exercising control over their data. Previous research has found that privacy labels can improve privacy knowledge (Kelley et al. 2009, 2010) and foster perceptions of trust and control (Xu et al. 2011; Pan and Zinkhan 2006). Furthermore, trust is positively associated with consumers' willingness to disclose personal information (Joinson et al. 2010), whereas privacy concern has the opposite influence (Culnan and Armstrong 1999). To overcome privacy concerns, it is important to build perceptions of privacy and control. In summary, we posit that the clear transparency enabled by the privacy label approach can serve to enhance privacy knowledge, build consumers' perceptions of privacy, control and trust, and enhance privacy self-efficacy, thus empowering consumers to utilise IoT devices while retaining some level of privacy. We argue that with this knowledge, consumers can choose what personal data to disclose to IoT devices.

7.7   Concluding Remarks

In the coming years, IoT is predicted to grow exponentially generating value for consumers in all aspects of their lives. Researchers have highlighted the importance of ensuring user privacy in the IoT context, stating users' privacy 'should be guaranteed' (Sicari et al. 2015, p. 151). Furthermore, as technology continues to increase in pervasiveness, it is important to explore how trust can be engendered in and between technologies that are built upon complex data exchange infrastructures and a lack of prior experience with the technology in question (van der Werff et al. 2018). In this chapter, we present a framework for examining the effectiveness of privacy disclosures on privacy and trust perceptions and consequently, enhancing adoption and sustained usage of IoT devices.

The framework is contextualised in the broad IoT context. Empirical research is needed to determine the effectiveness of the proposed privacy label and the framework itself in different IoT contexts, applications, and other dimensions. For example, adaptation may be required for use cases such as conversational assistants where data collection occurs verbally and may require the consideration of factors outlined by Saffarizadeh et al. (2017). Moreover, there is a need for research that maps out the privacy issues across the broader IoT landscape including the device, connection, and application layers discussed in Chap. 1.

In addition to addressing consumer perceptions regarding privacy and trust related to IoT, it is important to consider technical advances such as fog computing. Fog computing can facilitate the realisation of many new applications on IoT devices, while also reducing latency, enabling mobility, location awareness and heterogeneity (Alrawais et al. 2017). In terms of security, the computational power offered by fog computing combined with the devices and sensors of the IoT could provide enhanced security to minimise attacks. However, issues related to privacy and trust are likely to be complicated by advances in fog computing (Alrawais et al. 2017). Further research, may look to adapt this framework for fog computing and other advances in technology that have privacy implications, not least artificial intelligence.

References

Abomhara, Mohamed, and Geir M. Køien. 2014. *Security and Privacy in the Internet of Things: Current Status and Open Issues.* 2014 international conference on privacy and security in mobile systems (PRISMS), 1–8. IEEE.

Alrawais, Arwa, Abdulrahman Alhothaily, Chunqiang Hu, and Xiuzhen Cheng. 2017. Fog Computing for the Internet of Things: Security and Privacy Issues. *IEEE Internet Computing* 21 (2): 34–42.

Bélanger, France, and Robert E. Crossler. 2011. Privacy in the Digital Age: A Review of Information Privacy Research in Information Systems. *MIS Quarterly* 35 (4): 1017–1042.

———. 2019. Dealing with Digital Traces: Understanding Protective Behaviors on Mobile Devices. *The Journal of Strategic Information Systems* 28 (1): 34–49.

Bélanger, France, and Heng Xu. 2015. The Role of Information Systems Research in Shaping the Future of Information Privacy. *Information Systems Journal* 25 (6): 573–578.

Belanger, France, Janine S. Hiller, and Wanda J. Smith. 2002. Trustworthiness in Electronic Commerce: The Role of Privacy, Security, and Site Attributes. *The journal of strategic Information Systems* 11 (3–4): 245–270.

Chanson, Mathieu, Andreas Bogner, Dominik Bilgeri, Elgar Fleisch, and Felix Wortmann. 2019. Blockchain for the IoT: Privacy-Preserving Protection of Sensor Data. *Journal of the Association for Information Systems* 20 (9): 10.

Cisco. 2013a. Internet of Everything: A $4.6 Trillion Public-Sector Opportunity. https://www.cisco.com/c/dam/en_us/about/business-insights/docs/ioe-public-sector-vas-white-paper.pdf.

———. 2013b. Embracing the Internet of Everything to Capture Your Share of $14.4 Trillion. https://www.cisco.com/c/dam/en_us/about/business-insights/docs/ioe-economy-insights.pdf.

Crossler, R. E., & Belanger, F. (2019). Why Would I Use Location-Protective Settings on My Smartphone? Motivating Protective Behaviors and the Existence of the Privacy Knowledge–Belief Gap. *Information Systems Research* 30 (3), 995–1006.

Crossler, Robert E., and France Bélanger. 2017. *The Mobile Privacy-Security Knowledge Gap Model: Understanding Behaviors.* Proceedings of the Hawaii International Conference on System Sciences.

Culnan, Mary J. 1993. How Did They Get My Name?: An Exploratory Investigation of Consumer Attitudes Toward Secondary Information Use. *MIS Quarterly* 17: 341–363.

Culnan, Mary J., and Pamela K. Armstrong. 1999. Information Privacy Concerns, Procedural Fairness, and Impersonal Trust: An Empirical Investigation. *Organization Science* 10 (1): 104–115.

Daubert, Joerg, Alexander Wiesmaier, and Panayotis Kikiras. 2015. *A View on Privacy & Trust in IoT.* 2015 IEEE International Conference on Communication Workshop (ICCW), 2665–2670. IEEE.

Davies, Nigel, Nina Taft, Mahadev Satyanarayanan, Sarah Clinch, and Brandon Amos. 2016. *Privacy Mediators: Helping IoT Cross the Chasm.* Proceedings of the 17th International Workshop on Mobile Computing Systems and Applications, 39–44. ACM.

Dinev, Tamara, and Paul Hart. 2006. An Extended Privacy Calculus Model for e-Commerce Transactions. *Information Systems Research* 17 (1): 61–80.

Donaldson, Thomas, and Thomas W. Dunfee. 1994. Toward a Unified Conception of Business Ethics: Integrative Social Contracts Theory. *Academy of Management Review* 19 (2): 252–284.

Fox, Grace, Colin Tonge, Theo Lynn, and John Mooney. 2018. *Communicating Compliance: Developing a GDPR Privacy Label.* Proceedings of the 24th Americas Conference on Information Systems.

Gefen, David, Izak Benbasat, and Paula Pavlou. 2008. A Research Agenda for Trust in Online Environments. *Journal of Management Information Systems* 24 (4): 275–286.

Haller, Stephan, Stamatis Karnouskos, and Christoph Schroth. 2008. *The Internet of Things in an Enterprise Context.* Future Internet Symposium, 14–28. Berlin, Heidelberg: Springer.

Hsu, Chin-Lung, and Judy Chuan-Chuan Lin. 2016. An Empirical Examination of Consumer Adoption of Internet of Things Services: Network Externalities and Concern for Information Privacy Perspectives. *Computers in Human Behavior* 62: 516–527.

ICO. 2017. Privacy Notices, Transparency and Control. A Code of Practice on Communicating Privacy Information to Individuals. https://ico.org.uk/for-organisations/guide-to-dataprotection/privacy-notices-transparency-and-control/

Joinson, Adam N., Ulf-Dietrich Reips, Tom Buchanan, Carina B. Paine, and Schofield. 2010. Privacy, Trust, and Self-disclosure Online. *Human–Computer Interaction* 25 (1): 1–24.

Keith, Mark J., Jeffry S. Babb, Paul Benjamin Lowry, Christopher P. Furner, and Amjad Abdullat. 2015. The Role of Mobile-Computing Self-efficacy in Consumer Information Disclosure. *Information Systems Journal* 25 (6): 637–667.

Kelley, Patrick Gage, Joanna Bresee, Lorrie Faith Cranor, and Robert W. Reeder. 2009. *A Nutrition Label for Privacy.* Proceedings of the 5th Symposium on Usable Privacy and Security, 4. ACM.

Kelley, Patrick Gage, Lucian Cesca, Joanna Bresee, and Lorrie Faith Cranor. 2010. *Standardizing Privacy Notices: An Online Study of the Nutrition Label Approach.* Proceedings of the SIGCHI Conference on Human factors in Computing Systems, 1573–1582. ACM.

Kim, Min Sung, and Seongcheol Kim. 2018. Factors Influencing Willingness to Provide Personal Information for Personalized Recommendations. *Computers in Human Behavior* 88: 143–152.

Kim, Dongyeon, Kyuhong Park, Yongjin Park, and Jae-Hyeon Ahn. 2019. Willingness to Provide Personal Information: Perspective of Privacy Calculus in IoT Services. *Computers in Human Behavior* 92: 273–281.

Li, Yuan. 2011. Empirical Studies on Online Information Privacy Concerns: Literature Review and an Integrative Framework. *CAIS* 28: 28.

———. 2012. Theories in Online Information Privacy Research: A Critical Review and an Integrated Framework. *Decision Support Systems* 54 (1): 471–481.

Lowry, Paul Benjamin, Tamara Dinev, and Robert Willison. 2017. Why Security and Privacy Research Lies at the Centre of the Information Systems (IS) Artefact: Proposing a Bold Research Agenda. *European Journal of Information Systems* 26 (6): 546–563.

Marakhimov, Azizbek, and Jaehun Joo. 2017. Consumer Adaptation and Infusion of Wearable Devices for Healthcare. *Computers in Human Behavior* 76: 135–148.

McKnight, D. Harrison, Michelle Carter, Jason Bennett Thatcher, and Paul F. Clay. 2011. Trust in a Specific Technology: An Investigation of Its Components and Measures. *ACM Transactions on Management Information Systems (TMIS)* 2 (2): 12.

McKinsey Global Institute. 2015. The Internet of Things: Mapping the Value Beyond the Hype. McKinsey & Company. https://www.mckinsey.com/~/media/McKinsey/Industries/Technology%20Media%20and%20Telecommunications/High%20Tech/Our%20Insights/The%20Internet%20of%20Things%20The%20value%20of%20digitizing%20the%20physical%20world/Unlocking_the_potential_of_the_Internet_of_Things_Executive_summary.ashx.

Pan, Yue, and George M. Zinkhan. 2006. Exploring the Impact of Online Privacy Disclosures on Consumer Trust. *Journal of Retailing* 82 (4): 331–338.

Park, Yong Jin, Scott W. Campbell, and Nojin Kwak. 2012. Affect, Cognition and Reward: Predictors of Privacy Protection Online. *Computers in Human Behavior* 28 (3): 1019–1027.

Pike, S., M. Kelledy, and A. Gelnaw. 2017. Measuring US Privacy Sentiment: An IDC Special Report.

Saffarizadeh, Kambiz, Maheshwar Boodraj, and Tawfiq M. Alashoor. 2017. *Conversational Assistants: Investigating Privacy Concerns, Trust, and Self-disclosure.* Thirty Eighth International Conference on Information Systems, South Korea.

Sicari, Sabrina, Alessandra Rizzardi, Luigi Alfredo Grieco, and Alberto Coen-Porisini. 2015. Security, Privacy and Trust in Internet of Things: The Road Ahead. *Computer Networks* 76: 146–164.

Tucker, Catherine E. 2014. Social Networks, Personalized Advertising, and Privacy Controls. *Journal of Marketing Research* 51 (5): 546–562.

van der Werff, Lisa, Colette Real, and Theo Lynn. 2018. Individual Trust and the Internet. In *Trust*, ed. R. Searle, A. Nienaber, and S. Sitkin. Oxford, UK: Routledge.

van der Werff, Lisa, Grace Fox, Ieva Masevic, Vincent C. Emeakaroha, John P. Morrison, and Theo Lynn. 2019. Building Consumer Trust in the Cloud: An Experimental Analysis of the Cloud Trust Label Approach. *Journal of Cloud Computing* 8 (1): 6.

Williams, Meredydd, Jason R.C. Nurse, and Sadie Creese. 2019. Smartwatch Games: Encouraging Privacy-Protective Behaviour in a Longitudinal Study. *Computers in Human Behavior* 99: 38–54.

Xu, Heng, Tamara Dinev, Jeff Smith, and Paul Hart. 2011. Information Privacy Concerns: Linking Individual Perceptions with Institutional Privacy Assurances. *Journal of the Association for Information Systems* 12 (12): 1.

Ziegeldorf, Jan Henrik, Oscar Garcia Morchon, and Klaus Wehrle. 2014. Privacy in the Internet of Things: Threats and Challenges. *Security and Communication Networks* 7 (12): 2728–2742.

CHAPTER 8

Mapping the Business Value of the Internet of Things

Pierangelo Rosati and Theo Lynn

Abstract The impacts of enterprise investments in technological infrastructure for the Internet of Things (IoT) go well beyond the technical domain and require significant changes in an enterprise's operations, strategy and approach to market. This chapter presents a framework for mapping the business value of IoT investments which aims to support managers in their decision-making process by providing an overview of how specific resources need to be linked together in order to generate business value. The presented framework is also used as a point of reference for identifying current research gaps which may represent avenues for future research.

Keywords Internet of Things • Business value • Information technology • Strategy • Value mapping

P. Rosati (✉) • T. Lynn
Irish Institute of Digital Business, DCU Business School, Dublin, Ireland
e-mail: pierangelo.rosati@dcu.ie; theo.lynn@dcu.ie

© The Author(s) 2020
T. Lynn et al. (eds.), *The Cloud-to-Thing Continuum*, Palgrave
Studies in Digital Business & Enabling Technologies,
https://doi.org/10.1007/978-3-030-41110-7_8

141

8.1   Introduction

Digital transformation is causing a strategic shift in organisations. Driven by key enabling technologies—big data analytics, cloud computing, mobile and social technologies—IT spending has been on the increase worldwide and forecasts to reach $6 trillion by 2022 (IDC 2018). While each of these technologies generates business and economic value on their own, they create much greater benefits when combined in to innovative solutions such as the Internet of Things (IoT) (Rosati et al. 2017).

Haller et al. (2009, p. 15) define IoT as "*a world where physical objects are seamlessly integrated into the information network, and where the physical objects can become active participants in business processes. Services are available to interact with these 'smart objects' over the Internet, query their state and any information associated with them, taking into account security and privacy issues.*" Even though the idea of connecting physical objects to the digital worlds is not completely new, the decreasing cost of sensors and computing resources, improvements in computing power and network infrastructures, and the flexibility and agility provided by cloud computing have made it possible for organisations to operationalise large IoT solutions (Agrawal et al. 2011; Ji et al. 2012; Sagiroglu and Sinanc 2013).

While most of the academic discussion has focused on the technical aspects of IoT, it should be noted that it also generates significant business opportunities (Côrte-Real et al. 2019; Wolf et al. 2019). Recent studies have investigated how IoT impacts organisations' business model (e.g., Fleisch et al. 2015; Dijkman et al. 2015; Metallo et al. 2018; Wolf et al. 2019) and how specific aspects of IoT may affect business value creation and extraction (e.g., Karkouch et al. 2016; Côrte-Real et al. 2019). However, clear methodologies for mapping, and indeed measuring, the business value of IoT are still missing.

This chapter aims to fill this gap by introducing a framework for mapping the enterprise business value of IoT and exploring the main cost and value drivers associated with IoT investments. The remainder of this chapter is organised as follows. Next, we introduce the typical IoT architecture and provide some exemplar use cases. Then we introduce the proposed mapping framework and discuss the main cost and value components. Finally, we conclude the chapter with a discussion and avenues for future research.

8.2   THE INTERNET OF THINGS

As discussed in Chap. 1, the term "IoT" is often used as an umbrella term for describing various aspects related to the extension of the Internet in to the physical world through "smart" devices (Miorandi et al. 2012). From a business perspective, IoT can be seen as an innovative hybrid construct which consists of two elements, the "thing" and the digital service, that are strictly interconnected in order to generate value (Fleisch et al. 2015). These two elements are brought together through a complex, modular, multi-layered architecture similar to the one represented in Fig. 8.1.

In the device layer, sensors and actuators transform real-word events in to digital signals (Ji et al. 2012). The network layer provides the network structure that allows a high number of connected devices to send information securely and with low latency (ITU 2012). The support layer provides the main functions related to data processing while the application layer provides the user interface of specific IoT application (ITU 2012).

According to recent estimates, worldwide hardware and software IoT spending is projected to grow, from $726 billion in 2019 to $1 trillion in 2022 (IDC 2019), and IoT solutions are expected to generate $4.6 trillion in value for the public sector and another $14.4 trillion for the private

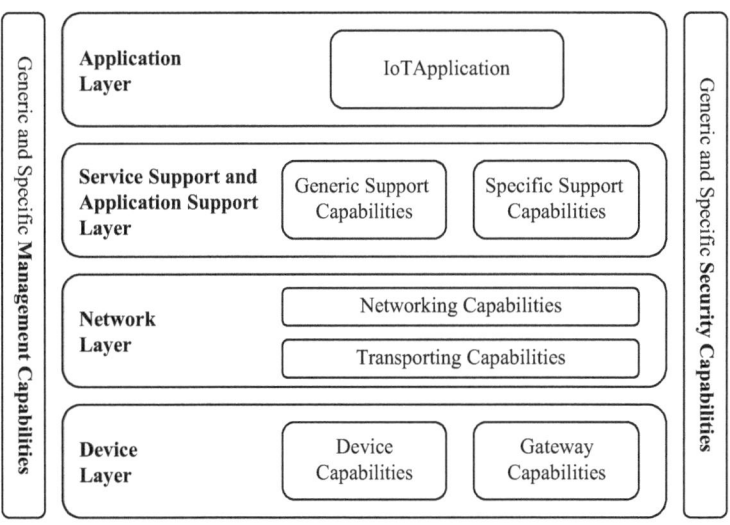

Fig. 8.1   IoT architecture. (Adapted from ITU 2012)

sector by 2023 (Cisco 2013a, b). Most of the value generated by IoT is linked to increased productivity and efficiency, cost reduction, better customer/citizen experience, faster innovation and new revenue streams (Cisco 2013a, b; Lynn et al. 2018).

Smart factories, for example, leverage data generated by different types of sensors to track the location of materials, machines, and other moveable assets in real time therefore enabling self-organising production lines, seamless synchronisation of production schedules and just-in-time supply chain management (Bansal 2019). The reduced number of people required to manage similar factories, lower inventories, and improved production quality could increase manufacturing efficiency by 27 per cent and add between $500 billion and $1.5 trillion in value to the global economy by 2022 (Capgemini 2017). Similarly, sensors embedded in a number of small connected devices enable remote monitoring of energy optimisation for smart buildings (King and Perry 2017).

Smart cities also represent a focus area with regard to IoT applications. In the context of smart cities, not only smart buildings, but also smart mobility and connected cars play a critical role (Singh 2016). Sensors can be used to map, for example, available parking slots throughout the city therefore reducing driving times and increasing the quality of life of citizens. Similarly, connected vehicles can provide better integration between different transport services and represent a breakthrough for the adoption of more efficient Mobility-as-a-Service solutions (Lennert et al. 2011; OECD 2019).

8.3   A VALUE MAPPING FRAMEWORK FOR THE INTERNET OF THINGS

While the technological infrastructure of IoT has attracted most of the attention from industry and academia (Del Giudice 2016), this is only part of the organisational ecosystem that enterprises have to build around their IoT offering. In this section, we discuss three parts of the IoT business value ecosystem—the value creators and generators, investment, and value generation and monetisation. We bring these together by adapting and extending Mikalef et al.'s (2019) value mapping framework for business data analytics.

8.3.1   *Value Creators and Consumers*

Mapping business value in IoT depends heavily on the perspective taken. The five main categories of actors for whom the IoT can generate value are (1) computing infrastructure providers, for example, hyperscale cloud service providers, (2) network infrastructure providers e.g. telecommunications companies, (3) application developers and providers (e.g. SAP), (4) device manufacturers or providers (e.g. Bosch or Apple), and (5) end users, whether organisations or consumers. Actors can play more than one role and indeed may create value together, for example, through co-creation. Regardless, to generate and capture the value from IoT, these actors must make investments.

8.3.2   *Investment*

Smart devices range from simple sensors with limited storage and processing power to relatively more advanced and complex devices such as smartphones. These devices are key enablers of big data as they generate constant streams of data that then get processed and analysed in order to return better services/products to the final user (Chen et al. 2012; Chanson et al. 2019). The expectation is that the IoT infrastructure enables these devices to interact with each other and with other systems with minimal latency regardless of their location or local computing power (Lynn et al. 2018). The traditional cloud computing paradigm where all the data is sent to a centralised (remote) cloud infrastructure, processed and sent back to the local device was not designed to meet the requirements of the IoT world.

The new infrastructure paradigm requires a continuum of computing resources activity from the cloud to the "thing" (C2T) where computing resources are located in the cloud, at the thing (edge computing), and/or somewhere in between (fog computing). As such, IoT is effectively driving the transformation of cloud computing in to a decentralised service architecture. Some of these new computing paradigms—fog computing, edge computing and dew computing are defined in Chap. 1. In this new technological landscape, the success or failure of an enterprise IoT service depends on the quality of the service provided by both cloud service providers and network operators, who have to decide where best to locate compute and storage resources along the cloud-to-thing continuum in order to meet Quality of

Service and Quality of Experience (QoE) requirements. The IoT assumes a multi-layered operational context with long IT value chains, where multiple actors have to work in sync in order to manage system complexity, while delivering an agreed QoE to the final user. In this context, the availability and quality of in-house IT "know-how" and "know-why" is almost as important as the "know-who" (Rennie 1999; Uden and He 2017). While the implications of IT investments on human capital and human resources available within the organisation are relatively simple to foresee, the strategic benefits of relational capital[1] (the "know who") or the resources required to improve it can be easily overlooked (Zardini et al. 2015).

The scale and complexity of data generated by smart end-points in the IoT is so complex, it is no longer realistic for IT teams to cost-effectively foresee and manage manually the infrastructure underlying the IoT or the data generated by the IoT on a detailed level due to high levels of dynamism and dependencies across the cloud-to-thing continuum (Domaschka et al. 2020). It is therefore necessary for enterprises to invest in organisation-wide analytics capability to realise value from the data generated by smart devices (Gupta and George 2016; Wamba et al. 2017) but also to manage the infrastructure and service chain underlying the IoT. As such, data analytics skills and resources are on the *must-have* list for enterprises that want to leverage IoT. However, given the high demand and scarcity of such resources, it is unsurprising that organisations are increasingly investing in algorithmic intelligence, one such example being AI/Ops—machine learning and artificial intelligence for IT operations (AI/Ops). It has also become evident that organisations must adopt a perspective that goes beyond the technical side when considering the effects and deployment of analytics (Mikalef et al. 2018). Skills and resources availability though is not enough to extract value from IoT. Enterprises have to find their way to combine all skills and resources in order to create unique capabilities which are aligned with the strategic objectives and allow them to adapt to the ever-changing competing landscape (Côrte-Real et al. 2019). Implementing a strategic approach to IoT and data investments and creating routines for faster development and deployment may represent key enablers of faster innovation and higher value creation.

---

[1] Relational capital can be defined as "all relationships—market relationships, power relationships and cooperation—established between firms, institutions and people, which stem from a strong sense of belonging and a highly developed capacity of cooperation typical of culturally similar people and institutions" (Capello and Faggian 2005, p. 75).

8.3.3   *Value Generation and Monetisation*

Irrespective of whether an enterprise adopts IoT for serving internal or external customers, the investment is justified only if the value generated exceeds the investment required. For a comprehensive investment evaluation of IoT though, the concept of value needs to be expanded to include the total value generated for all stakeholders (e.g. investors, employees, customers and suppliers) (Wolf et al. 2019). According to Haller et al. (2009), there are two main sources where enterprises can generate business value from the IoT: real world visibility and business process decomposition. Real word visibility is related to the fact that IoT bridges the gap between the physical and the digital words. In so doing, IoT provides enterprises with real-time insights in to what is happening in the real world thus enabling more effective optimisation, and better decision making. Business process decomposition relates to the fact the distributed nature of the IoT infrastructure enables more decentralised business processes therefore increased scalability, performance, and innovation.

Unfortunately, creating value is not enough. In an enterprise context, value is only relevant when captured and somehow monetised (Osterwalder et al. 2005; Zott et al. 2011), either directly (tangible benefits) or indirectly (intangible benefits). Hui (2014) provides a comparison between the main drivers of value creation and value capture of traditional and IoT products (Table 8.1).

Table 8.1   Mindset for the IoT industry. (Adapted from Hui 2014)

Value creation	*Traditional product mindset*	*IoT mindset*
Customer needs	Solve for existing needs and lifestyle in a reactive manner	Address real-time and emergent needs in a predictive manner
Offering	Standalone product that becomes obsolete over time	Product refreshes through over-the-air updates and has synergy value
Role of data	Single point data is used for future product requirements	Information convergence creates the experience for current products and enables services
Value capture		
Path to profit	Sell the next product or device	Enable recurring revenue and increase efficiency
Control points	Potentially includes commodity advantages, IP ownership, and brand	Adds personalisation and context; network effects between products
Capability development	Leverage core competencies, existing resources and processes	Understand how other ecosystem partners make money

With regard to value creation, data and information availability and faster time-to-market clearly play a central role. Valuable IoT solutions combine and integrate "thing"-based functions with IT-based functions (Fleisch et al. 2017) and separate the information flow generated by the device from its physical attributes (Wolf et al. 2019). As such, information (data) generates value in itself as it enables data-driven service innovation, real-time interactions with customers and to better predict future or emerging customer needs (Bohli et al. 2009; Bucherer and Uckelmann 2011; Holler et al. 2015). The IT-based function of IoT products is arguably the one that adds most value and the fact that each device is connected to the network, and to each other, allows providers to implement updates and introduce additional features "over the air". This dramatically reduce the time-to-market of innovation and it is particularly important for physical products with a longer lifecycle than software. A typical example comes from the automotive industry where Tesla Motors introduced the Autopilot function in to tens of thousands of cars already sold overnight through a software update (Kessler and Buck 2017).

IoT solutions also enable the creation of digital platforms where multiple actors can benefit from the unprecedented amount of information generated by connected devices. The value of information is non-exhaustive. On the contrary, it increases with use (Bohli et al. 2009). The amount of information available in the IoT world makes it the perfect environment for the nurturing inter-organisational collaborations and innovation, and for leveraging network effects which would ultimately benefit all stakeholders (Mejtoft 2011). IoT also allows enterprises to shift from unit-based revenue streams to value-based pricing which are more flexible and based on the value of service and information provided to the final user (Kindström 2010). However, this implies that the service provider is able to measure the value parameter associated with specific services and provide customers with transparent and clear value proposition (Kindström 2010).

Finally, IoT investments should ultimately provide the enterprise with the basis for creating a competitive advantage. This is easier said than done; in the IoT world, success lays on the edge between open innovation and collaboration, and internal knowledge management (Santoro et al. 2018). This is even more challenging when the competitive and technological landscape is in a constant state of change. Teece et al. (1997, p. 5) suggest that organisations should develop their dynamic capabilities "*to integrate, build, and reconfigure internal and external competences to address rapidly changing environments*". However, potential configurations are contingent upon the specific environment in which an organisation operates (Pavlou and El Sawy 2010). This implies that managers should develop and evaluate their own strategic approach to IoT as the impacts of technology investments permeate the entire business model and go well beyond the technical components.

8.3.4 Bring It All Together: A Value Mapping Framework for the Internet of Things

Figure 8.2 provides a graphical overview of our value mapping framework for the Internet of Things adapted an extended from Mikalef et al. (2019) for IoT. This framework is specifically designed to support managers when assessing the value of complex IoT investments.

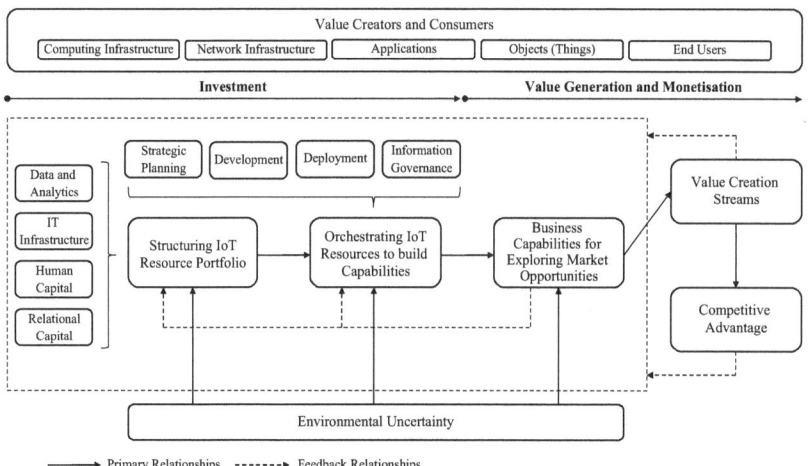

Fig. 8.2 Value mapping framework for the Internet of Things. (Adapted and extended from Mikalef et al. 2019)

8.4   Towards a Research Agenda on the Business Value of IoT

The value mapping framework presented in Fig. 8.2. Table 8.1 can also be used as a point of reference for developing a research agenda on different aspects of the business value of IoT. As IoT is situated at the intersection of a number of technologies, this research agenda may present avenues for future research across a number of disciplines such as information systems, computer science, and management.

IoT is a key enabler of Big Data analytics as sensors allow enterprises to collect constant streams of data of various types to obtain real-time insights into the real world. While data is mostly perceived as a valuable asset. However, storing massive volumes of data may have significant implications from a business and IT perspective; data is only valuable as long as it generates business outcomes (Sivarajah et al. 2017). This last step is not always straightforward and in fact organisations tends to adopt more of a deductive than an inductive approach to analytics projects (Constantiou and Kallinikos 2015). It can be hard sometimes to understand what kind of insights can be extracted from a specific type of data ex-ante. In the context of IoT, sensors need be embedded in to physical products/devices and the temptation to include different kind of sensors can be high as it may be the difficult to add them post-sale/installation. These sensors may generate streams of data that may remain unused and have regulatory and cost implications. In contrast, if a sensor that is able to capture valuable data is missing from the device, this may generate significant loss in revenues and/or costs post-sale for updates or replacements. Future research may clarify what are the benefits and challenges associated with inductive or deductive approaches toward analytics and potentially develop guidelines for IoT data monetisation.

From an infrastructure perspective, IoT introduces significant complexity mostly due to the fact that it requires a number of actors (e.g., cloud providers, cloud carriers, cloud brokers, edge device producers, etc.) to work together towards the same goal, that is, a seamless user experience. Some of the technological investments made by one actor may impact positively or negatively the others involved in the IoT value chain and therefore generate unexpected outcomes. A comprehensive evaluation of potential investments and dependencies in the wider IoT infrastructure needs to be carefully evaluated in the context of a longer service chain, and consequently wider value chain, rather than at an organisational level.

Future research may investigate expected or unexpected value transfer with the IoT supply chain as a consequence of IT investments therefore providing useful insights with regard to supply chain value creation.

Human and relational capital represent the "softer" side of the IoT resource portfolio. Sousa and Rocha (2019) identify three main groups of skills that organisations need to have in order to create digital businesses, that is, innovation skills, leadership skills and management skills. While the need for further skills development in IT-oriented contexts is widely recognised, there is still a need for clear guidelines on how to develop such skills and how to update them over time in order to meet the ever-changing market requirements (Sousa and Rocha 2019). Relational capital consists of relations that the company creates with different stakeholders. This is particularly important in the context of IoT which is mostly characterised by high competition, a complex value chain and low switching costs. Future research may investigate how relational capital is built in this context and hot it translates in business outcomes.

Resource orchestration is a key element for extracting value for IoT solutions. Collecting appropriate resources is not enough to be successful in the IoT world. Enterprises need to develop organisation-wide capabilities for leveraging IoT resources (Gupta and George 2016; Mikalef et al. 2019) but how this can be achieved is still unclear. Future research may provide organisations with a framework for developing such capabilities over time and map out potential enablers and constraining forces.

Once value it has been created, enterprises should be able to capture and somehow measure it. Value capturing can be mostly related to suitable business models for IoT. While previous studies have looked at the high-level impact of IoT on business models (Hui 2014; Dijkman et al. 2015; Fleisch et al. 2015; Metallo et al. 2018), future research may delve into each of the key elements of business models and explore different options available to enterprises to monetise their IoT solutions. Finally, enterprises should be able to measure the value generated by IoT investments. Enterprises are profit-driven organisations where value is typically measured in monetary terms. Over time, a number of methodologies to estimate the financial value generated by IT investments have been developed (see Table 8.2).

The length, complexity and opacity of the chain of service provision in the IoT may make the quantitative measurement of IoT business value extremely difficult, not least establishing causal relationships. Measuring the business value of technology investments can be considered both a

Table 8.2   Selected financial metrics for measuring the business value of IT investments

Metric	Description
Cost-benefit analysis (CBA)	CBA compares costs to benefits and therefore represents a measure of efficiency.
Economic value added (EVA)	EVA measures the value generated by an investment net of all costs including the cost of the capital invested. When evaluating different investment opportunities with similar expected returns, managers should opt for the one that generates the highest EVA.
Internal rate of return (IRR)	IRR represents the discount rate that would return a value of zero for NPV. Financially valuable investments have an IRR that is equal or higher than the desired or minimum rate of return.
Net present value (NPV)	NPV is a measure of the present value of the future cash flows generated by an investment, net of the initial capital outlay and discounted by a rate that reflects the time value of money and the risk of the investment.
Payback period	Payback period measures the time needed for a project to repay the initial investment. Investments with a shorter payback period may be more attractive however this metric does not provide any indication about the value generated after the payback period and therefore investment decisions should not be based on this metric alone.
Return on investment (ROI)	ROI is an accounting ratio that compares the net benefit generated by an investment to the overall investment required. As such, it allows to directly compare investments of different scale.
Total cost of ownership (TCO)	This metric captures the overall cost of single components of an IT system such as hardware, software, maintenance etc. TCO is relatively simple to calculate but it does not capture the benefits the system generates to the organisation and therefore it only provides an incomplete picture of the overall investment.

science and an art (Tallon et al. 2020). In fact, the use of technology in enterprises today is so widespread and pervasive that the impacts of IT investments typically go beyond tangible operational benefits and costs to include organisational and business impacts (e.g., increased agility, faster innovation, better employees or customer experience) that are intangible in nature and therefore hard to quantify (Tallon and Kraemer 2007; Tallon et al. 2020; Rosati and Lynn 2020). This is unlikely to change for IoT and, indeed, may be exacerbated. The increasing adoption of IT "as-a-Service" makes it easier to forecast, monitor and quantify operational costs (Rosati et al. 2017). This leaves managers with more time to evaluate potential intangible impacts of the investment. Unfortunately, there is no *one-size-fits-all* methodology for doing this. Tallon (2014) proposes a *distributed sensemaking* model where managers, in the absence of objective data, rely on the views of multiple internal stakeholders to notice, weigh, and filter informational cues from various sources in order to reach a reasoned, balanced judgment of the intangible value delivered by the IT investment. Future research may map out tangible and intangible costs and benefits generated by different types of IT investments for different actors along the IoT value chain, and provide guidelines to measuring the business value of IoT.

8.5 Conclusion

In this chapter, we presented a business value mapping framework for the Internet of Things with purpose of identifying the main actors, cost and value drivers associated with IoT. The scale, interconnectivity and complexity of the Internet of Things makes conceptualising and measuring business value extremely challenging. Despite this, given the opportunity and risks, it is essential. Building on Mikalef et al. (2019), we provide a preliminary framework for mapping business value in the IoT that can be used by enterprises to identify areas for strategic investment and consideration in this exciting new space.

References

Agrawal, D., S. Das, and A. El Abbadi. 2011. *Big Data and Cloud Computing: Current State and Future Opportunities.* Proceedings of the 14th International Conference on Extending Database Technology, 530–533. ACM.

Bansal, N. 2019. This is How a Smart Factory Actually Works. *World Economic Forum.* https://www.weforum.org/agenda/2019/06/connectivity-is-driving-a-revolution-in-manufacturing/.

Bohli, J.M., C. Sorge, and D. Westhoff. 2009. Initial Observations on Economics, Pricing, and Penetration of the Internet of Things Market. *ACM SIGCOMM Computer Communication Review* 39 (2): 50–55.

Bucherer, E., and D. Uckelmann. 2011. Business Models for the Internet of Things. In *Architecting the Internet of Things*, 253–277. Berlin, Heidelberg: Springer.

Capello, R., and A. Faggian. 2005. Collective Learning and Relational Capital in Local Innovation Processes. *Regional Studies* 39 (1): 75–87.

Capgemini. 2017. Smart Factories: How can Manufacturers Realize the Potential of Digital Industrial Revolution. https://www.capgemini.com/wp-content/uploads/2017/05/dti-smart-factories-full-report-rebranded-web-version_16032018.pdf.

Chanson, M., A. Bogner, D. Bilgeri, E. Fleisch, and F. Wortmann. 2019. Blockchain for the IoT: Privacy-Preserving Protection of Sensor Data. *Journal of the Association for Information Systems* 20 (9): 10.

Chen, H., R.H. Chiang, and V.C. Storey. 2012. Business Intelligence and Analytics: From Big Data to Big Impact. *MIS Quarterly* 36 (4): 1165–1188.

Cisco. 2013a. Embracing the Internet of Everything to Capture Your Share of $14.4 Trillion. https://www.cisco.com/c/dam/en_us/about/ac79/docs/innov/IoE_Economy.pdf.

———. 2013b. Internet of Everything: A $4.6 Trillion Public-Sector Opportunity. https://www.cisco.com/c/dam/en_us/about/business-insights/docs/ioe-public-sector-vas-white-paper.pdf.

Constantiou, I.D., and J. Kallinikos. 2015. New Games, New Rules: Big Data and the Changing Context of Strategy. *Journal of Information Technology* 30 (1): 44–57.

Côrte-Real, N., P. Ruivo, and T. Oliveira. 2019. Leveraging Internet of Things and Big Data Analytics Initiatives in European and American Firms: Is Data Quality a Way to Extract Business Value? *Information & Management* 57: 103141.

Del Giudice, M. (Ed.). (2016). Discovering the Internet of Things (IoT): technology and business process management, inside and outside the innovative firms. Emerald.

Dijkman, R.M., B. Sprenkels, T. Peeters, and A. Janssen. 2015. Business Models for the Internet of Things. *International Journal of Information Management* 35 (6): 672–678.

Domaschka, Jorg, Frank Griesinger, Mark Leznik, Per-Olov Östberg, Keith Ellis, Paolo Casari, Frank Fowley, and Theo Lynn. 2020. Towards an Architecture for Reliable Capacity Provisioning for Distributed Clouds. In *Managing Distributed Cloud Applications and Infrastructure: A Self-Optimising Approach.* Cham: Springer.

Fleisch, E., M. Weinberger, and F. Wortmann. 2015. Business Models and the Internet of Things. In *Interoperability and Open-Source Solutions for the Internet of Things*, 6–10. Cham: Springer.

———. 2017. Geschäftsmodelle im Internet der Dinge. In *Industrie 4.0*, 1–16. Wiesbaden: Springer Vieweg.

Gupta, M., and J.F. George. 2016. Toward the Development of a Big Data Analytics Capability. *Information & Management* 53 (8): 1049–1064.

Haller, S., A. Karnouskos, and C. Schroth. 2009. The Internet of Things in an Enterprise Context. In *Future Internet Symposium*, 14–28. Berlin, Heidelberg: Springer.

Holler, J., V. Tsiatsis, C. Mulligan, S. Avesand, S. Karnouskos, and D. Boyle. 2015. *From Machine-to-Machine to the Internet of Things*. Elsevier.

Hui, G. 2014. How the Internet of Things Changes Business Models. *Harvard Business Review* 92 (7/8): 1–5.

IDC. 2018. ICT Spending Forecast—2018–2022 Forecast. https://www.idc. com/promo/global-ict-spending/forecast.

———. 2019. IDC Forecasts Worldwide Spending on the Internet of Things to Reach $745 Billion in 2019, Led by the Manufacturing, Consumer, Transportation, and Utilities Sectors. https://www.idc.com/getdoc.jsp?contai nerId=prUS44596319.

ITU. 2012. Overview of the Internet of Things. http://handle.itu. int/11.1002/1000/11559.

Ji, C., Y. Li, W. Qiu, U. Awada, and K. Li. 2012. *Big Data Processing in Cloud Computing Environments*. 2012 12th International Symposium on Pervasive Systems, Algorithms and Networks, 17–23. IEEE.

Karkouch, A., H. Mousannif, H. Al Moatassime, and T. Noel. 2016. Data Quality in Internet of Things: A State-of-the-Art Survey. *Journal of Network and Computer Applications* 73: 57–81.

Kessler, T., and C. Buck. 2017. How Digitization Affects Mobility and the Business Models of Automotive OEMs. In *Phantom Ex Machina*, 107–118. Cham: Springer.

Kindström, D. 2010. Towards a Service-based Business Model–Key Aspects for Future Competitive Advantage. *European Management Journal* 28 (6): 479–490.

King, J., and C. Perry. 2017. Smart Buildings: Using Smart Technology to Save Energy in Existing Buildings. https://aceee.org/sites/default/files/publica tions/researchreports/a1701.pdf.

Lennert, F., C. Macharis, V. van Acker, and L. Neckermann. 2011. Smart Mobility and Services—Expert Group Report. https://ec.europa.eu/transparency/reg-expert/index.cfm?do=groupDetail.groupDetailDoc&id=34596&no=1.

Lynn, T., P. Rosati, and P.T. Endo. 2018. Toward the Intelligent Internet of Everything: Observations on Multidisciplinary Challenges in Intelligent Systems Research. Coloquio Doctorados: Tecnología, Ciencia y Cultura: una visión globa. Puebla, Mexico..

Mejtoft, T. 2011. *Internet of Things and Co-creation of Value*. 2011 International Conference on Internet of Things and 4th International Conference on Cyber, Physical and Social Computing, 672–677. IEEE.

Metallo, C., R. Agrifoglio, F. Schiavone, and J. Mueller. 2018. Understanding Business Model in the Internet of Things Industry. *Technological Forecasting and Social Change* 136: 298–306.

Mikalef, P., I.O. Pappas, J. Krogstie, and M. Giannakos. 2018. Big Data Analytics Capabilities: A Systematic Literature Review and Research Agenda. *Information Systems and e-Business Management* 16 (3): 547–578.

Mikalef, P., I.O. Pappas, J. Krogstie, and P.A. Pavlou. 2019. Big Data and Business Analytics: A Research Agenda for Realizing Business Value. *Information & Management*.

Miorandi, D., Sicari, S., De Pellegrini, F., & Chlamtac, I. (2012). Internet of things: Vision, applications and research challenges. *Ad hoc networks, 10*(7), 1497–1516.

OECD. 2019. Enhancing the Contribution of Digitalisation to Smart Cities of the Future. https://www.oecd.org/cfe/regional-policy/Smart-Cities-FINAL.pdf.

Osterwalder, A., Y. Pigneur, and C.L. Tucci. 2005. Clarifying Business Models: Origins, Present, and Future of the Concept. *Communications of the Association for Information Systems* 16 (1): 1.

Pavlou, P.A., and O.A. El Sawy. 2010. The "Third Hand": IT-Enabled Competitive Advantage in Turbulence Through Improvisational Capabilities. *Information Systems Research* 21 (3): 443–471.

Rennie, M. 1999. Accounting for Knowledge Assets: Do We Need a New Financial Statement? *International Journal of Technology Management* 18 (5–8): 648–659.

Rosati, P., and T. Lynn. 2020. Measuring the Business Value of Infrastructure Migration to the Cloud. In *Measuring the Business Value of Cloud Computing*, ed. T. Lynn, J.G. Mooney, P. Rosati, and G. Fox. Springer.

Rosati, P., G. Fox, D. Kenny, and T. Lynn. 2017. *Quantifying the Financial Value of Cloud Investments: A Systematic Literature Review.* 2017 IEEE International Conference on Cloud Computing Technology and Science (CloudCom), 194–201. IEEE.

Sagiroglu, S., and D. Sinanc. 2013. *Big Data: A Review.* 2013 International Conference on Collaboration Technologies and Systems (CTS), 42–47. IEEE.

Santoro, G., D. Vrontis, A. Thrassou, and L. Dezi. 2018. The Internet of Things: Building a Knowledge Management System for Open Innovation and Knowledge Management Capacity. *Technological Forecasting and Social Change* 136: 347–354.

Singh, K. 2016. Contribution on Smart Urban Mobility for Safe, Inclusive, Resilient and Sustainable Cities. https://www.iru.org/sites/default/files/2016-10/0354_UN_Habitat_report_web-FINAL.pdf.

Sivarajah, U., M.M. Kamal, Z. Irani, and V. Weerakkody. 2017. Critical Analysis of Big Data Challenges and Analytical Methods. *Journal of Business Research* 70: 263–286.

Sousa, M.J., and Á. Rocha. 2019. Skills for Disruptive Digital Business. *Journal of Business Research* 94: 257–263.

Tallon, P.P. 2014. Do You See What I See? The Search for Consensus Among Executives' Perceptions of IT Business Value. *European Journal of Information Systems* 23 (3): 306–325.

Tallon, P.P., and K.L. Kraemer. 2007. Fact or Fiction? A Sensemaking Perspective on the Reality Behind Executives' Perceptions of IT Business Value. *Journal of Management Information Systems* 24 (1): 13–54.

Tallon, P.P., J.G. Mooney, and M. Duddek. 2020. Measuring the Business Value of IT. In *Measuring the Business Value of Cloud Computing*. Springer.

Teece, D.J., G. Pisano, and A. Shuen. 1997. Dynamic Capabilities and Strategic Management. *Strategic Management Journal* 18 (7): 509–533.

Uden, L., and W. He. 2017. How the Internet of Things can Help Knowledge Management: A Case Study from the Automotive Domain. *Journal of Knowledge Management* 21 (1): 57–70.

Wamba, S.F., A. Gunasekaran, S. Akter, S.J.F. Ren, R. Dubey, and S.J. Childe. 2017. Big Data Analytics and Firm Performance: Effects of Dynamic Capabilities. *Journal of Business Research* 70: 356–365.

Wolf, V., J. Stumpf-Wollersheim, and L. Schott. 2019. The Internet of Things in a Business Context: Implications with Respect to Value Creation, Value Drivers, and Value Capturing. In *Digital Entrepreneurship*, 185–197. Cham: Springer.

Zardini, A., F. Ricciardi, and C. Rossignoli. 2015. The Relational Capital of the IT Department: Measuring a Key Resource for Creating Strategic Value. *Journal of Intellectual Capital* 16 (4): 835–859.

Zott, C., R. Amit, and L. Massa. 2011. The Business Model: Recent Developments and Future Research. *Journal of Management* 37 (4): 1019–1042.

INDEX